D1203306

Machu Picchu

A Civil Engineering Marvel

by
Kenneth R. Wright, P.E., and
Alfredo Valencia Zegarra, Ph.D.

with
Ruth M. Wright, J.D., and
Gordon McEwan, Ph.D.

American Society of Civil Engineers
1801 Alexander Bell Drive
Reston, Virginia 20191-4400

Abstract: This volume presents a new perspective into the Lost City of the Inca based on science and engineering. The geography, site selection, engineering infrastructure, and the city planning are described with site-specific photography and narrative. The authors provide detailed analyses of the Inca water system, drainage, agriculture, stonework, and construction methods. The background on the comprehensive technology transfer system utilized by the Inca engineers is given in terms of their cultural heritage. An amply illustrated walking guide and detailed map makes Machu Picchu and its environs a reality for engineers and non-engineers alike.

Library of Congress Cataloging-in-Publication Data

Wright, Kenneth R.

Machu Picchu : a civil engineering marvel / by Kenneth R. Wright and Alfredo Valencia Zegarra ... [et al.].

p. cm.

Includes bibliographical references.

ISBN 0-7844-0444-5

1. Civil engineering—Peru—Machu Picchu Site. 2. Hydraulic engineering—Peru—Machu Picchu Site. 3. Incas—Antiquities. 4. Machu Picchu Site (Peru). I. Valencia Zegarra, Alfredo. II. Title.

TA52 .W75 2000
624'0985'37—dc21 00-024452

Library of Congress Catalog Card No: 00-024452
ISBN 0-7844-0444-5
Manufactured in the United States of America

Dedication

This book is dedicated to many individuals and two organizations. Each played an important role in our engineering, exploration, and technical research at Machu Picchu.

Hiram Bingham, a Yale history professor, was an explorer extraordinaire who brought Machu Picchu to the attention of the world after he came across the ruins high in the Andes in 1911. When Bingham returned to Machu Picchu in 1912 with a team of engineers, surveyors, and scientists, he meticulously cleared the site and photographed the remarkable ruins. These photographs appeared in the April 1913 issue of the *National Geographic Magazine*. In his announcement to the world about the Lost City of the Inca, Bingham noted that "the Inca were good engineers."

Many years have passed since the Bingham 1911 discovery of Machu Picchu, and during those years nearly every school child has become aware of this ancient royal estate high in the Andes. However, it has not been until now that his conclusion about the Inca being good engineers has been probed and defined in specific terms. Our 1994–1999 research has proven Bingham correct. In exploring the civil engineering achievements of the Inca at Machu Picchu, we developed a respect for Bingham's excellent field work and his documentation. We thank Yale University and the National Geographic Society for sponsoring his exploration nearly a century ago.

U.S. Senator Timothy Wirth assisted Kenneth Wright in the early stages of the Machu Picchu engineering project by lending encouragement and providing an introduction to Peruvian officials prior to his becoming the Secretary of State for Global Affairs. Without Senator

Wirth's assistance there likely would have been no archaeological permit for the paleohydrological team.

Professor Frank Eddy of the University of Colorado Anthropology Department was instrumental in the initiation of the research by lending encouragement and advising us of the merits of the proposed paleohydrological project.

Professor John Rowe and his colleague Pat Lyon of the University of California at Berkeley provided encouragement and insight for the technical effort. Their vast knowledge of the Inca Empire, and Machu Picchu in particular, helped us prove correct Bingham's statement that the Inca were good engineers.

Our heartfelt thanks go to Professor Richard Burger and Lucy Burger of the Yale Peabody Museum for lending their support to our project and for providing unqualified permission to use the historic Bingham photographs. We met the Burgers amongst the ruins of Machu Picchu during the research effort.

The Instituto Nacional de Cultura of Lima and Cusco provided excellent assistance via several directors and longtime archaeologist Señorita Arminda Gibaja Oviedo, director of the Cusco Regional Museum. Then there was Señorita Elva Torres, who discovered the gold bracelet in Test Pit No. 6 while collecting soil samples for us. We also dedicate this book to the numerous local Quechua Indians who cleared many kilometers of rain forest trails and provided Quechua language prayers for the success of our work.

Finally, we dedicate this book to the entire staff of Wright Water Engineers, who have assisted in the field exploration and the necessary office technical support.

Kenneth R. Wright and Alfredo Valencia Zegarra

Contents

Preface

Machu Picchu, the royal estate of the Inca ruler Pachacuti, is the most well-known of all Inca archaeological sites. While numerous visitors view the site in awe each year, little is known about its hydrogeology and paleohydrology and how the various components of the engineering infrastructure of this Andean community functioned. For this reason, the Peruvian government issued an archaeological permit to us in 1994. The objectives of our research were to define the ancient water use and water handling engineering of these pre-Columbian Native Americans.

The primary author, Kenneth R. Wright, is with the consulting firm of Wright Water Engineers of Denver. He has a long interest in engineering achievements by ancient people and their use and handling of water. His interest in Machu Picchu began in 1974, when his wife returned from an archaeological trip to Peru and asked how Machu Picchu, situated on a ridge top, could have gotten its water supply. For many years he sought an exploration permit, and in 1994 the Instituto National de Cultura authorized his research.

Professor Alfredo Valencia Zegarra, Ph.D., teaches at the Universidad de San Antonio Abad in Cusco, Peru. After five years as resident archaeologist at Machu Picchu, he continued his interest in, and research there. He is recognized as a leading archeologist on the Inca achievements at Machu Picchu. Dr. Valencia knows each nook and cranny of this royal estate.

Dr. Gordon McEwan is a specialist in New World archaeology. He has served at the Dumbarton Oaks Museum and as curator at the Denver Art Museum. He presently is associate professor at Wagner College in New York. Since receiving his Ph.D. from the University of Texas, he

has focused on significant field excavations and research at the Wari Administrative Center of Pikillacta and nearby Chokepukio, both in the Cusco area. Dr. McEwan also has served as a consultant to Time-Life on the Inca Empire. Dr. McEwan contributed Chapter 9 to this book, which tells us how the Inca were able to achieve so much in such a short time.

Ruth Wright, an attorney, award-winning photographer, and former Colorado state legislator, is due full credit for being the spark plug on the engineering research at Machu Picchu. Upon returning from Machu Picchu in 1974, she asked the simple question, "from where did the Inca engineers derive their water?" She graciously contributed Chapter 10, which provides a civil engineer's walking tour.

Many others from the Machu Picchu paleohydrological team have contributed to this technical analysis of Machu Picchu.

The research methods included common civil engineering methods, extensive field investigations, field instrument surveying, flow measurements, and geologic evaluations. Reviews of published material on Machu Picchu and interviews with Andean scholars were essential to be able to conceptualize the ancient engineering achievements within the Inca cultural framework. Once in the field, networking with local officials and Inca archeological experts proved most helpful.

Our original paleohydrological objectives were achieved early on—in 1996. As a result, the research was expanded to include the agricultural practices at Machu Picchu, stonework, construction methods, exploration, and the preparation of a detailed map. The exploration focused on the lower east flank of Machu Picchu that had lain under a dense rain forest for four and a half centuries.

This book is intended to generally summarize for civil engineers our experiences and findings at Machu Picchu. Chapters 1 through 8 provide detailed discussions of various civil engineering features at Machu Picchu, such as site constraints and planning and construction issues. Chapter 9, by Dr. Gordon McEwan, lends cultural and archeological context to the rest of the book. Chapter 10 is a walking tour of Machu Picchu, written by Ruth Wright. It will lead you to most of the features discussed throughout the book. To accompany Chapter 10, you may wish to have the full-color, enhanced map of Machu Picchu. If so, you can receive a complimentary map from the authors by contacting Kenneth R. Wright at Wright Water Engineers, Inc., 2490 W. 26th Avenue, Suite 100A, Denver, CO 80211.

Chapter One

Mystical Machu Picchu

The Lost City of the Inca

Machu Picchu was built by Native Americans before the arrival of the Spanish Conquistadors. It was abandoned after the Inca Empire collapsed and endured under a thick rain forest until the 20th century. Scientists, engineers, and laymen alike continue to marvel at the wonders of Machu Picchu and its magical ambiance. The mystique of Machu Picchu is in its details: the hydrology of the water supply, the hydraulics of the canal and fountains, and the blending of man's work with the challenging natural topography and environment with which the civil engineers were faced.

One cannot fully appreciate the accomplishment that Machu Picchu represents without first considering the site constraints its ancient civil engineers faced. The Inca engineering feats of planning and construction must be viewed in context with its setting, geology, and climate. When these details are viewed as a whole, Machu Picchu becomes the Lost City of the Inca in all its glory.

The Ancient Royal Estate

Machu Picchu, the royal estate of the Inca ruler Pachacuti (Rowe 1990), is a breathtaking monument to the ancient engineering skills of the Inca people (Figures 1-1 and 1-2). Construction of Machu Picchu began in A.D. 1450. It was burned in A.D. 1562 and finally abandoned 10 years later (Maurtua 1906; Rowe 1990,1997). However, it likely ceased normal operation by A.D. 1540 due to the collapse of the Inca empire.

Although the Inca did not have a written language, the well-preserved remains of Machu Picchu show that they had an advanced understanding of such principles as urban planning, hydrology, hydraulics, drainage, and durable construction methods. By studying the Inca's engineering techniques, in conjunction with the natural environment, we are able to supplement existing archaeological theory on the Inca's practices, religion, and the significance of Machu Picchu (Figure 1-3).

Machu Picchu's technical planning is surely the key to the site's longevity and functionality. The Inca's careful use of hydraulic, drainage, and construction techniques ensured that the retreat was not reduced to rubble during its many years of abandonment. These techniques, combined with a strong knowledge of hydrology, were what made it a grand and operational retreat high in the most rugged of terrain.

The rugged terrain, the steep (sometimes vertical) cliffs, and the slippery-when-wet soils formed from the parent granite rock meant that construction could not be performed without risk to human life.

FIGURE 1-1►
Machu Picchu, built on a ridgetop at an elevation of 2,440 meters (8,000 feet), presented many engineering challenges during its construction. Machu Picchu is often cloud-shrouded. This view from the Inca Trail from Cusco was an Inca traveler's first view of Machu Picchu.

FIGURE 1-2▼
The terraces of Machu Picchu have endured for five centuries, representing a tribute to the technical abilities of the Inca civil engineers. Terraces, stairways, and buildings exist even near the top of Huayna Picchu, the high peak to the right.

One only needs to view the terraces, stairways, and buildings high up near the top of the Huayna Picchu (Figure 1-2) to envision a workman's long fall to the river below.

The Setting and Geology

Machu Picchu is about 1,400 kilometers (870 miles) south of the Equator on the eastern slope of the Peruvian Andes (Figure 1-4). The site lies near the headwaters of the Amazon River, at longitude 72° 32′ west and latitude 13° 9′ south. Machu Picchu is laid out like a patchwork quilt on a mountain ridge between two prominent mountain peaks—Machu Picchu and Huayna Picchu (Figure 1-5). The dramatic location, 500 meters (1,640 feet) above the valley bottom, is a result of tectonic forces and valley downcutting by the Urubamba River that meanders along three sides of the ridge-top retreat. Further away, but visually dominating, are the ice-capped peaks of Mt. Veronica to the east

◀FIGURE 1-3
The Intiwatana, atop a high natural pyramid, is known as the Hitching Post of the Sun. However, it was not a solar observatory. The stonework was carefully shaped without the use of iron or steel.

(5,850 meters or 19,159 feet) and Mt. Salcantay to the south (6,257 meters or 20,530 feet).

This Inca site lies within the Cordillera Oriental (Eastern Cordillera) between the high plateau and subandine zones of the Peruvian Andes (Marocco 1977) on a 40-square-kilometer (15 square mile) portion of the complex Vilcabamba Batholith (Caillaux n.d.). This 250-million-year-old intrusion is white-to-gray-colored granite, characterized by its abundance of quartz, feldspar, and mica (predominantly biotite). This mineralogical composition made the granite of Machu Picchu a durable construction material. It was the inherent, rectangular joint pattern of the granite that the Inca workmen masterfully utilized in building stones to construct Machu Picchu that ensured the retreat's longevity.

The most significant geologic characteristics of the Machu Picchu site are the numerous faults and abundant rock fractures. The two principal faults are named for the two prominent local peaks: Huayna Picchu fault and Machu Picchu fault, as shown in Figure 1-6. These high-angle reverse faults formed a wedge-shaped structural block that dropped relative to the peaks on either side. This block, or graben, is the structure on which the ancient Inca people built their retreat. The Machu Picchu fault influences the location at which the Machu Picchu spring emerges. Increased permeability along upgradient

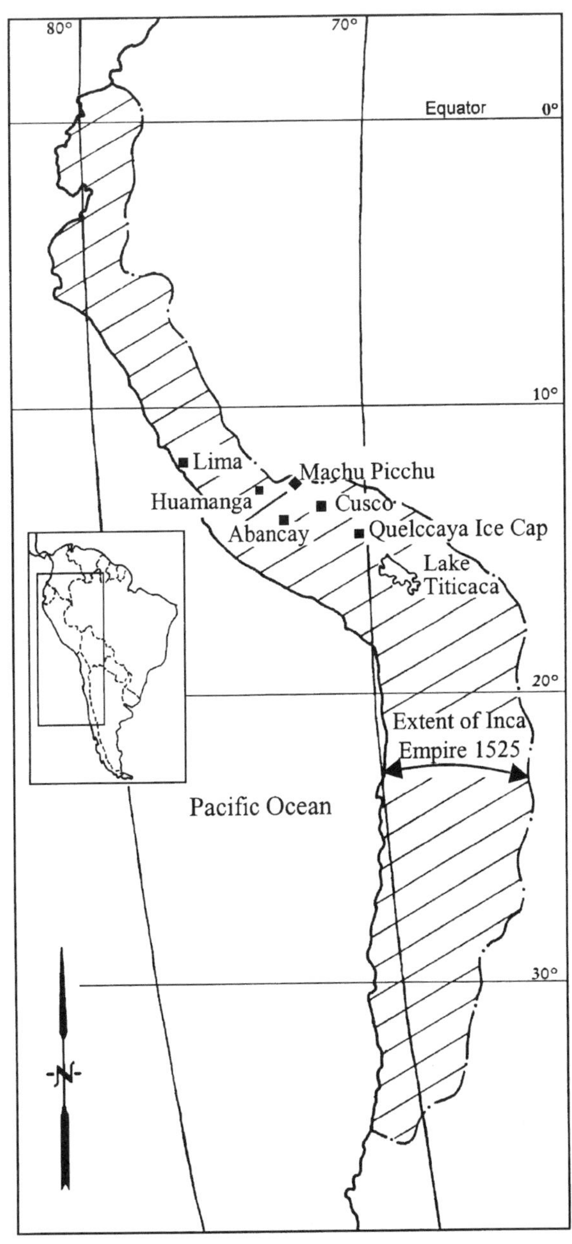

FIGURE 1-4▶
This is a map of the Inca Empire at the time of the arrival of the Spanish Conquistador Pizarro in 1532. Machu Picchu is located 80 kilometers (50 miles) northwest of the Inca Empire capital of Cusco. Lima is shown only to provide orientation.

▲FIGURE 1-5
The view from Machu Picchu Mountain provides a good perspective of the patchwork quilt-like layout of Machu Picchu. The encircling Urubamba River and the holy mountain of Huayna Picchu are in the background.

portions of the fault system allows precipitation to infiltrate and then emanate at the spring site, thus providing a perennial water source to the Inca people.

The Machu Picchu fault system is also responsible for much of the topographic relief in the vicinity of the mountain sanctuary. The orientation of the fault system can be identified by aligning the near-vertical northwestern face of Machu Picchu with the linear reach of the nearby Urubamba River along the southeastern flank of Mount Putucusi.

Ancient Climate

The ancient climate affecting Machu Picchu had two distinct phases. The first phase, from A.D. 1450 to A.D. 1500, was characterized by precipitation likely averaging about 1,830 millimeters (72 inches) per year, while the second period, from A.D. 1500 to A.D. 1540, was characterized by an estimated 2,090 millimeters (82 inches) per year of annual rainfall. The estimates of precipitation for these two phases are based on our comparison of modern climatological data collected at the site and the ice core data collected from the Quelccaya Ice Cap (Thompson et al. 1989). This ice core research provides a unique look back into the climatological record and is a result of work by the Byrd Polar Research Center (NGDC, NOAA 1986).

The Quelccaya Ice Cap is 250 kilometers (155 miles) southeast of Machu Picchu in the Cordillera Blanca mountain range of southern Peru. The ice cap coring research shows various periods of abundant precipitation and frequent dry periods as shown in Figure 1-7. Clima-

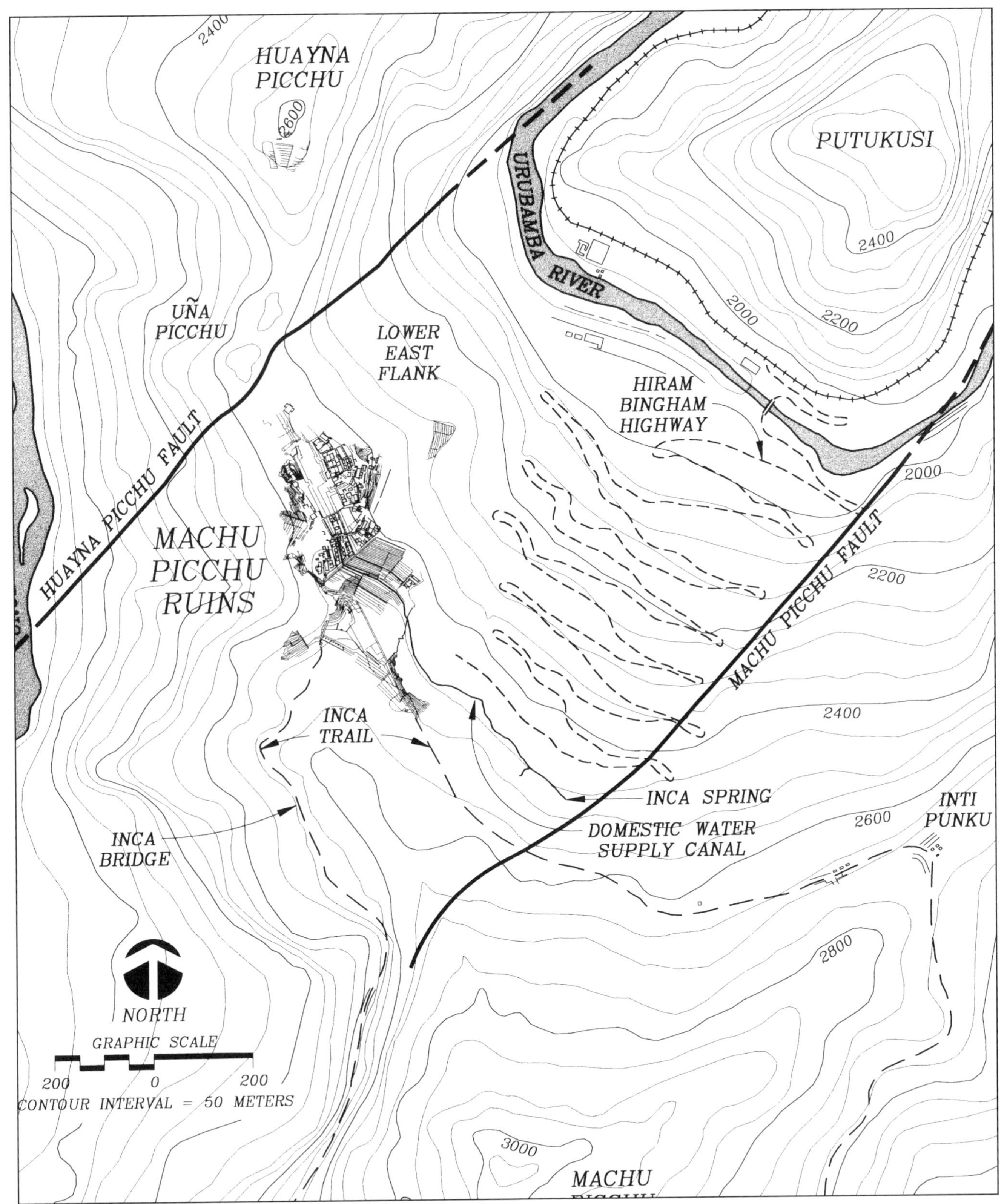

▲FIGURE 1-6

The high angle reverse faults of Huayna Picchu and Machu Picchu resulted in the down-drop of the area in between. The fracturing of the rock near the faults helped create thousands of raw building blocks. Rock fracturing also provided the geologic basis for the primary spring that lies downhill from the main Inca Trail to Cusco.

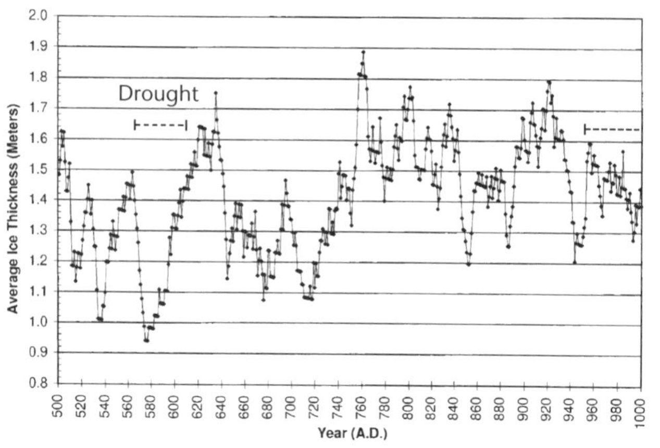

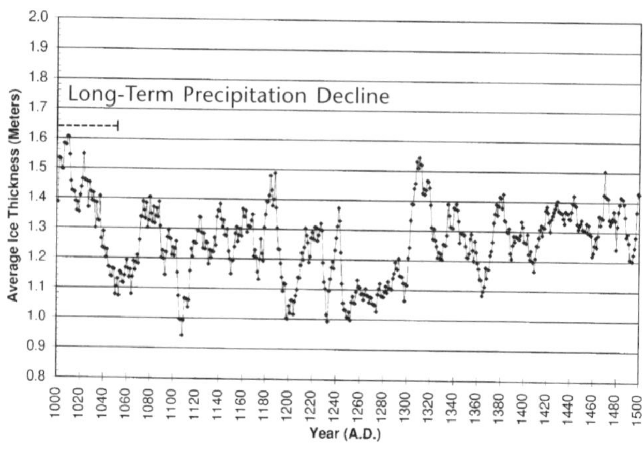

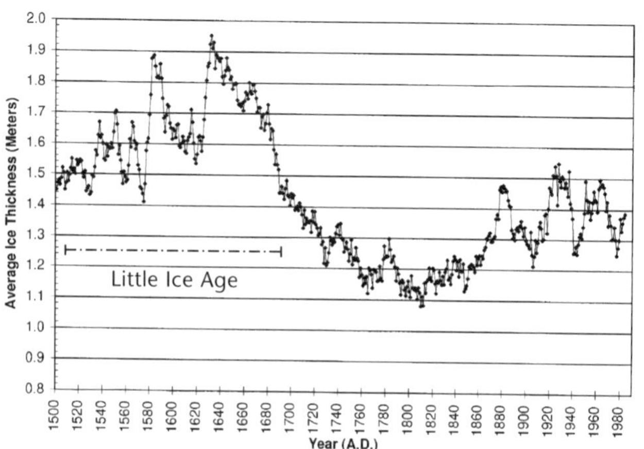

FIGURE 1-7

The climate of the Southern Andes was etched in ice. The 1,500-year-record, defined by Ohio State University, tells us much about the ancient precipitation patterns. The 32-year drought late in the 6th century was devastating to the Tiwanaku Empire, as was the precipitation decline from A.D. 950 to 1050. The Little Ice Age, commencing in A.D. 1500, is well-known among paleoclimatologists.

tologists and glaciologists may take special interest in the 200-year-long "Little Ice Age," which began in A.D. 1500 (Thompson et al. 1986). The ice core data evaluation indicates a long-term (A.D. 470 to A.D. 1984) average ice accumulation of about 1.4 meters (4.6 feet) per year and a similar modern period accumulation per year (Thompson et al. 1989).

The estimated decadal precipitation for the 90-year period of primary occupation of Machu Picchu is presented in Table 1-1. Modern climatic data were collected at Machu Picchu from 1964 to 1977 and indicate an average annual precipitation of 1,960 millimeters (77 inches). The average precipitation for 1964 through 1977 is shown for comparison (Wright et al. 1997).

TABLE 1-1
Annual Precipitation at Quelccaya by Decade Relative to Modern Precipitation

Decade	*Equivalent Annual Precipitation (mm/yr)*
1450–59	1770
1460–69	1900
1470–79	1830
1480–89	1770
1490–99	1860
1500–09	2020
1510–19	2150
1520–29	1980
1530–39	2220
1964–1977	1960

Site Selection

The Machu Picchu site is extraordinary; modern visitors describe it as breathtaking (Figure 1-8). The Inca planners and engineers judged the site to be well suited for a royal estate. First, it is surrounded on three

◄FIGURE 1-8
When viewed from near the Temple of Three Windows, the eastern urban sector displays its terraces and walls with a variety of angular lines and shadows. The long granite staircase provides easy access between levels.

FIGURE 1-9►
The rounded peak of Putucusi (left) is still revered by the descendants of the Inca, along with Huayna Picchu and Machu Picchu Mountains. A hang glider is in the center of the photograph.

sides by the steep, roaring Urubamba River. Second, the site is on a graben between two regional faults that created the two sharp peaks of the Huayna Picchu and Machu Picchu Mountains. Across the river is the steep Putucusi peak, which has a rounded profile like a half orange when viewed from Machu Picchu (Figure 1-9). All three peaks, even today, are considered by the local Quechua Indians to be holy mountains. Further away are the triangular-shaped Yanantin mountain and the glacier-capped flanks of Mt. Veronica, as illustrated in Figures 1-10 and 1-11. Mt. Salcantay, at an elevation of 6,257 meters (20,530 feet), is some 20 kilometers (12 miles) south. The arrow stone in Figure 1-12 on the Intiwatana summit points toward Mt. Salcantay.

FIGURE 1-10▲
Yanantin Mountain is an imposing triangular-shaped high peak toward which many of the views from Machu Picchu are oriented. Putucusi is the dark rounded peak in the lower right.

But without the perennial spring on the north slope of Machu Picchu Mountain, there would have been no Machu Picchu to admire in the early morning (Figure 1-13). Prior to the emperor's selection of the site for his royal estate, the civil engineers would have had to determine that the water supply would be suitable. Pachacuti was one to be reckoned with, as evidenced by Figure 1-14, and the engineers had to be sure of their water supply evaluations.

◀FIGURE 1-11
The glacier-covered ridges of Mount Veronica provide an imposing setting for the 5,850 meter (19,159 feet) summit as viewed from the top of Huayna Picchu.

▲FIGURE 1-12
This south-pointing arrow stone is situated at the summit of the Intiwatana of Machu Picchu. It is similar to three others at Machu Picchu, all pointing toward Salcantay Mountain some 20 kilometers (12 miles) away.

▲FIGURE 1-13
Early morning mists are common at Machu Picchu. Here, the Intiwatana is in the center, Uña Picchu peak behind, and the shadow of Huayna Picchu is behind and to the right.

FIGURE 1-14▶
This 16th-century drawing by Don Felipe Huamán Poma de Ayala shows the Inca Emperor Pachacuti ready for battle while holding two traditional Inca tools of warfare and a shield for protection (Poma de Ayala 1978).

Chapter Two

Engineering Planning

Once the Inca emperor approved the Machu Picchu site for his royal estate, the civil engineers had to further analyze the ridge top characteristics. They already knew there was a reliable spring, that the dense rain forest meant that irrigation was unnecessary, and that there was an abundance of good, fractured rock for building purposes. Because the emperor wanted great views of mountains and the river, special structures would need to be oriented to capture the views.

The site was in a transitional zone between Cusco, the empire's base of power, and the untamed Amazon basin, so the Inca would rely on the Urubamba River and cliffs encircling the site on three sides for natural defenses.

While the selected site was steep and subject to landslides, the abundance of fractured rock meant that many terraces could be built.

City Planning

After the thickly forested site of Machu Picchu was cleared, the civil engineers saw that an agriculture area could be laid out separate and apart from an urban area to the north (Figure 2-1). For security, an outer wall would be constructed incorporating the natural topography, and an inner wall between the agricultural and urban areas would have only one gate to serve the Inca Trail from the capital city of Cusco (Figure 2-2). Near the city, on the Inca Trail, a three-sided Guardhouse (Figure 2-3) would control access, with a *kallanka* for housing local visitors. There would also be a narrow branch trail. Security for this branch trail would be supplemented by the use of a drawbridge-like structure on the steep cliff of the Machu Picchu fault (Figure 2-4).

The Sacred (Main) Gate, the one gate through the inner wall, would be oriented so as visitors passed through the gate, they would surely be impressed by a framed view of the holy Huayna Picchu Mountain (Figure 2-5). Storehouses and llama pens just inside the gate, which could be closed, would facilitate the delivery and unloading of goods.

Consistent with Inca city planning practices there would be upper and lower sectors in the urban area, with a large plaza in between. This would require intensive site work, massive walls, and substantial filling to create flat building sites and the central plaza (Figure 2-6).

The major temples and the Royal Residence would be located in the upper (western) urban sector. Fortunately, the water of the peren-

FIGURE 2-1▶
The huge agricultural sector lies to the south of the urban sector, separated by the long east-west Main Drain. The Inca Trail from Cusco can barely be seen in the upper left. The northerly urban sector is in the foreground.

◀FIGURE 2-2
The one and only gate into Machu Picchu is shown in the lower left. Two narrow openings are situated beyond the gate for penning llama, the beasts of burden, while across the roadway are warehouses for foodstuff delivery and storage.

FIGURE 2-3▲
Huayna Picchu and the three-sided Guardhouse are seen here with the Ceremonial Rock in the foreground. The Guardhouse is situated at the junction of two trails approaching Machu Picchu.

◄FIGURE 2-4
A branch of the Inca Trail crossed the sheer slip face of the Machu Picchu fault on a stone revetment constructed with a moveable log crossing. The drawbridge-like design provided increased security to Machu Picchu.

FIGURE 2-5►
The one gate into Machu Picchu was carefully designed to provide visitors with an impressive framed view of Huayna Picchu Mountain.

FIGURE 2-6►
The layout of Machu Picchu included an eastern and western (upper and lower) sector separated by a large plaza running through the middle. This view from Machu Picchu Mountain looks north. The rugged trail to Huayna Picchu can be seen in the center of the photograph.

nial spring on the north slope of Machu Picchu Mountain could be carried by gravity flow to the center of the upper urban sector as long as the canal grade was carefully controlled. The terminus of the canal would discharge water into the first fountain (Figure 2-7). The Royal Residence would be adjacent to this fountain. A large granite rock protruding near Fountain No. 1 nearby would be the site for the important Temple of the Sun (Figure 2-8). A series of 16 fountains paralleling a long stairway (Figure 2-9) would continue down the steep slope, serving both the upper and lower urban sectors with a domestic water supply. The Royal Residence would be elegant and private; there would be only one entrance into the royal quarters and it would be well-drained and secure. The perennial spring and the canal grade dictated the locations of Fountain No. 1 and the Royal Residence.

The ridge top itself would be the logical place for the main temples, as it afforded incomparable views. Here would be a sacred plaza

FIGURE 2-7▶
Fountain No. 1 included a huge stone slab into which a rectangular basin was cut. An orifice outlet was connected to a smoothly shaped channel that carried the water to the lower fountains. The Inca emperor used this uppermost fountain.

FIGURE 2-8▲
The Temple of the Sun houses an important carved natural rock huaca *that is related to an eastern window, the two serving as a solar observatory. The* huaca *is visible through the Enigmatic Window above the two Quechua Indians.*

with the Temple of Three Windows on the east, the Principal Temple on the north, an overlook to the west, and a service building on the south (Figure 2-10). The very highest point of the Machu Picchu urban area would be reserved for the Intiwatana. This pyramid-like rock would need numerous terraces to ward off topsoil erosion and the stonecutters would need to flatten the pinnacle-like top to create a platform. While doing so, they would carve the very top of the natural rock into a rectangular prism for use during special religious ceremonies. Then, below the summit, the stonecutters would carve another natural granite outcrop to emulate the holy mountains of Yanantin and Putucusi, which dominate the view from the granite staircase leading to the summit (Figure 2-11).

The more ordinary buildings such as housing and additional storehouses would be located in the lower (eastern) urban sector, but so would a series of shrines to the south to be used by the residents and elite visitors (Figure 2-12). Directly across the plaza from the Intiwatana would be three double-jamb doorways accessed by a long stairway and flat paths and guarded by two sentry posts. Then, to the south, rough granite rocks that tended to look somewhat like wings would become the Temple of the Condor, the "wings" enhanced with additional stone walls (Figure 2-13). Beneath, would be a rock carved to represent the head of a condor, complete with a ruff.

Overlooking the valley on the east was a rock cave. Here, a solar observatory would be con-

◄FIGURE 2-9
The Stairway of the Fountains is shown descending from the area of the Temple of the Sun, while paralleling a series of 16 domestic water supply fountains. Fountain No. 10 at the upper left is the only fountain with an east–west jet of water. Fountain No. 16 is at the far right, behind the stone wall.

◄FIGURE 2-10
The Sacred Plaza is formed by the Temple of Three Windows on the left, a service building (upper), the Principal Temple toward the camera, and a viewing platform to the right.

◄FIGURE 2-11
To help the Inca demonstrate their power over the land and water, they often cut image stones in the shape of important mountains beyond. The view to the east is from the granite stairway leading to the Intiwatana, a focal point for Machu Picchu.

FIGURE 2-12▲
Machu Picchu has many shrines and "stones of adoration" at locations that provide spectacular views of rugged mountains.

FIGURE 2-13►
An integration of natural granite rocks and carefully constructed stone walls provide the appearance of huge wings at the Temple of the Condor. Many "New-Age" visitors find the Temple of the Condor especially important because of the underground caverns, the niches, and the various rooms.

structed so for a few days during the December solstice, the rising sun's rays would penetrate deep into the cave.

To the north, nearer the holy mountain of Huayna Picchu, a grouping of rock pinnacles jutted upwards. In time, it could become an important temple. Farther on, and closer to Huayna Picchu, a sacred rock would be shaped to the image of the mountain ranges beyond (Figure 2-14). Finally, the summit of Huayna Picchu would be made accessible via a carefully built mountain trail with granite stairways, similar to the trail that would lead to the summit of Machu Picchu Mountain (Figure 2-15). The upper slopes and summit of Huayna

◄FIGURE 2-14
The Sacred Rock, flanked by two wayronas, is shaped to imitate Yanantin Mountain in the background. Some investigators believe, however, that the Sacred Rock better reflects the mountain range in the opposite direction. Dr. Alfredo Valencia Zegarra excavated and restored this area in 1968 (Valencia 1977).

◄FIGURE 2-15
Near the summit of Huayna Picchu Mountain, this granite stairway provides an easy climb to the top.

FIGURE 2-16▲
This helicopter photograph of Machu Picchu, looking westerly, illustrates how the Inca engineers fit the enclave to the ridge top. The outer wall is seen in the left center extending downward to the right. The Main Drain separates the agricultural area from the urban area.

Picchu would be the site of elaborate terraces, small temple-like buildings, and a stone arrow pointing south toward Salcantay, many kilometers (miles) away—one of the most revered mountains of the Inca.

All of this building development on the Machu Picchu ridge would generally be connected with horizontal north–south paths and steep east–west stairways. All paths and stairs would fit into an overall plan for security, convenient passages, and respect for the limited access to the more royal upper urban sector where checkpoints and sentry stations would be planned. This overall plan, as it exists today, is shown in Figure 2-16. When Yale University professor Hiram Bingham rediscovered Machu Picchu in 1911 and cleared the site in 1912, he found Machu Picchu in much the same shape it is in today. His 1912 photograph of the Sacred Plaza and the Intiwatana, provided from the National Geographic archives, is displayed in Figure 2-17.

Engineering Infrastructure

The grand plan for Machu Picchu also included longevity, for without a timeless life of Machu Picchu, the emperor would not be well served. The civil engineers of Machu Picchu knew from the technology gleaned from preceding empires and other conquered peoples stretch-

FIGURE 2-17▲
In 1912, Yale professor Hiram Bingham took this glass plate photograph of the Sacred Plaza with the Intiwatana in the background. The Principal Temple facing the camera and the Temple of Three Windows on the right existed then, just as they do today. The photograph was provided by National Geographic Magazine.

ing from north to south, the importance of a sound and well-conceived infrastructure.

The infrastructure would need to rely on an incorporated knowledge of hydrology, hydraulics, agriculture, urban drainage, sanitation, soils and foundation technology, structural engineering, and a "storehouse" of construction methods along with good survey control of elevations, distances, and alignment. The following chapters describe these inspiring Inca technical achievements.

Chapter Three

Hydrology

The Inca civil engineers had an uncanny skill in the field of water development. No matter where they built throughout their 4,200-kilometer-long (2,600-mile-long) empire, the Inca seemed able to capitalize on minor water sources or to harness and manage great rivers. Somehow, the Inca understood the rudiments of hydrology along with the seasonal variations of water flow without the use of piezometers, slide rules, or even a written language. They used this skill at Machu Picchu.

Location and Water

A high mountain ridge between two prominent peaks seems an unlikely location for ancient people to have found a pure and reliable groundwater source. The Inca engineers would not have built the royal estate of Machu Picchu at this location if they had not found and developed the perennial spring on the steep north slope of Machu Picchu Mountain (Figure 3-1). Nearly 2,000 millimeters (79 inches) of annual rainfall, a modestly sized tributary drainage basin, igneous bedrock, and extensive faulting collectively provided Pachacuti and his engineers with a reliable domestic source of water. This water supply made it feasible for the Inca ruler to build his sanctuary at this spectacular location. Machu Picchu supported a resident population of about 300 people with a maximum of 1,000 for nearly a century, from A.D. 1450 to about A.D. 1540. The site was finally abandoned in A.D. 1572 (Rowe 1990).

As chapter 4 describes, the Inca hydrologists and hydraulic engineers were responsible for establishing the location of the focal point of Machu Picchu—The Sun Temple and the emperor's residence (Figure 3-2). The emperor needed to have the first, or uppermost, fountain close at hand.

Hydrogeology of the Main Water Source

To better understand the Inca's ancient engineering practices, the authors conducted detailed studies of the Machu Picchu spring hydrology (Wright et al. 1997b). The elevation of the steep drainage basin

tributary to the Inca springs ranges from 2,458 meters (8064 feet) to 3,050 meters (10,007 feet), a vertical rise of 592 meters (1,940 feet). This tributary drainage basin covers 16.3 hectares (40 acres), including two subbasins of 5.9 (15 acres) and 10.4 hectares (26 acres). After detailed field inspections, site observations, photographic interpretation, and topographic mapping, we delineated the drainage basins. The drainage basin is well-covered with tropical forest vegetation and bisected by the final section of the ancient Inca Trail from Intipunku at the ridge top to the royal estate of Machu Picchu (Figure 1-6).

FIGURE 3-1▲
The watershed of the Inca spring is shown in the background on the north slope of Machu Picchu Mountain. The Inca Canal lies on a terrace just above the upper storehouse.

We prepared an inflow–outflow evaluation for the 10.4-hectare (26 acres) topographic drainage basin using the observed annual rainfall of 1,960 millimeters (77 inches), an annual evapotranspiration of the forest cover estimate of 1,760 millimeters (69 inches), and the approximate yield of the spring of 40,000 cubic meters (32 acre feet) per year. We assumed the spring was 100 percent efficient and used an average daily evapotranspiration of 4.82 millimeters (0.19 inches) (Wright 1997b). This evaluation helped determine whether the Machu Picchu fault was responsible for a hydrogeologic zone of capture greater than that defined by the topographic ground surface drainage basin. The evaluation concluded that the yield of the primary spring represents drainage from a hydrogeologic catchment basin as much as twice as large as the topographic drainage basin. This tends to be consistent with our field surveys made on the north slope of the Machu Picchu Mountain and our review of topographic mapping.

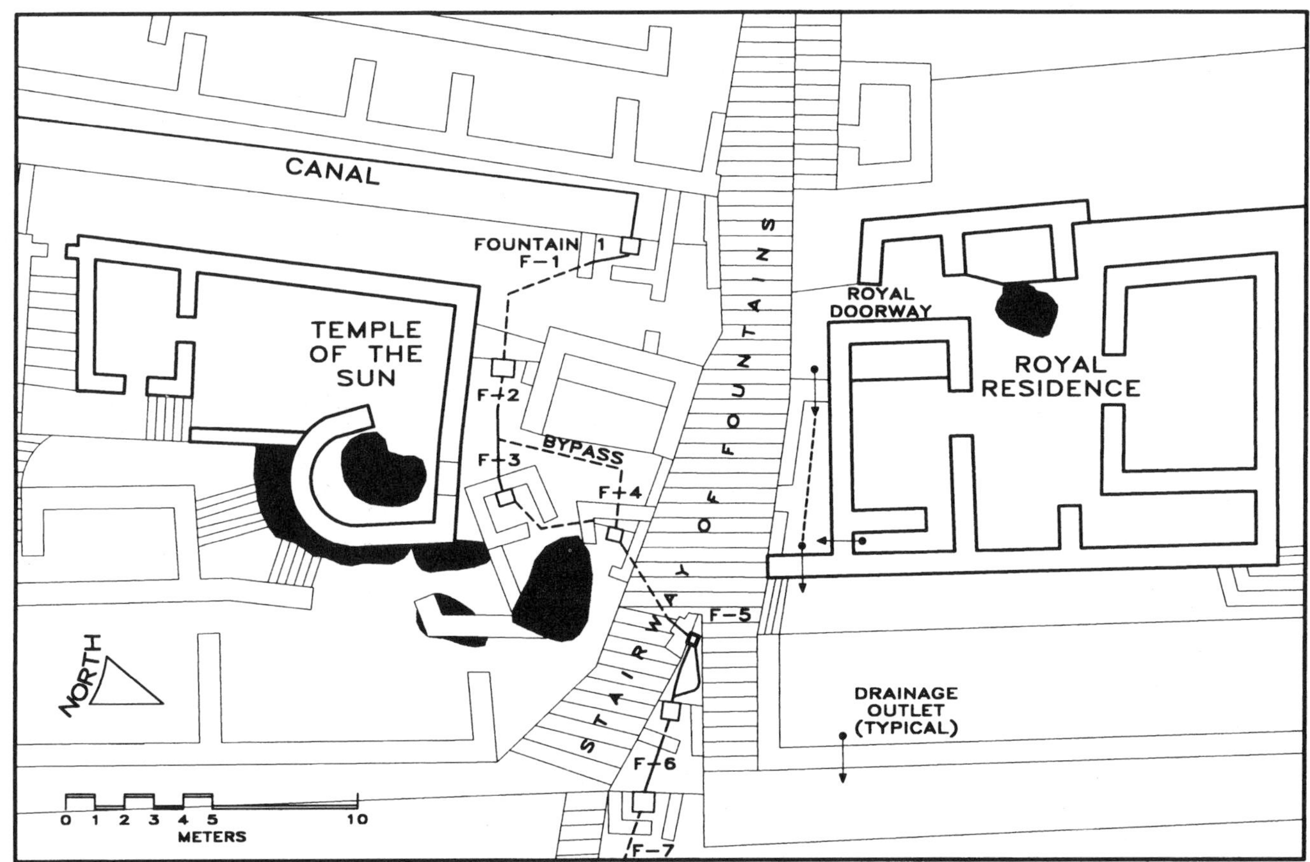

FIGURE 3-2▲
While the elevation of the spring controlled the location of Fountain No. 1, it also controlled the siting of the Royal Residence so the emperor would have first use of the water from Fountain No. 1. Fountain Nos. 1–7 are shown, along with the drainage for the emperor's one and only doorway.

Since Hiram Bingham's "discovery" of Machu Picchu in 1911, some scientists have speculated that the Inca abandoned the royal estate due to a water shortage. This theory has no merit when the ice core data from the Quelccaya glacier are analyzed (Wright, Dracup, Kelly 1996). The ice core data suggest that the "Little Ice Age" began in about A.D. 1500, and the precipitation for the final decade of occupancy was greater than any of the other decades of the occupancy period (Table 1-1) (Wright et al. 1997).

Spring Collection Works

While the Machu Picchu primary spring is a natural phenomenon, its reliable yield is enhanced by an innovative and well-engineered stone spring collection system that still functions today (Figure 3-3). Located on the north slope of Machu Picchu Mountain at an elevation of 2,458 meters (8,060 feet), this ancient example of groundwater and hydraulic engineering is no simple spring works. Rather, it is a carefully planned and built permeable stone wall set into the steep hillside. The linear stone wall is approximately 14.6 meters (48 feet) long and up to 1.4 meters (4.6 feet) high. At the foot of the wall is a rectangular collection trench with a cross-section approximately 0.8 meter (2.6 feet) wide and about 0.6 meter (2 feet) high. The spring works were accessed for operation and maintenance by a terrace approximately 1.5 to 2 meters (5 to 7 feet) wide, supported on the steep hillside by a stone wall along the full length of the collection area. A

FIGURE 3-3►
The well-engineered Inca spring still stands today on the steep 50 percent slope of Machu Picchu Mountain. Water percolates through the stone wall on the left into the channel in the center. Hydrogeologist Gary Witt is shown recording flow data.

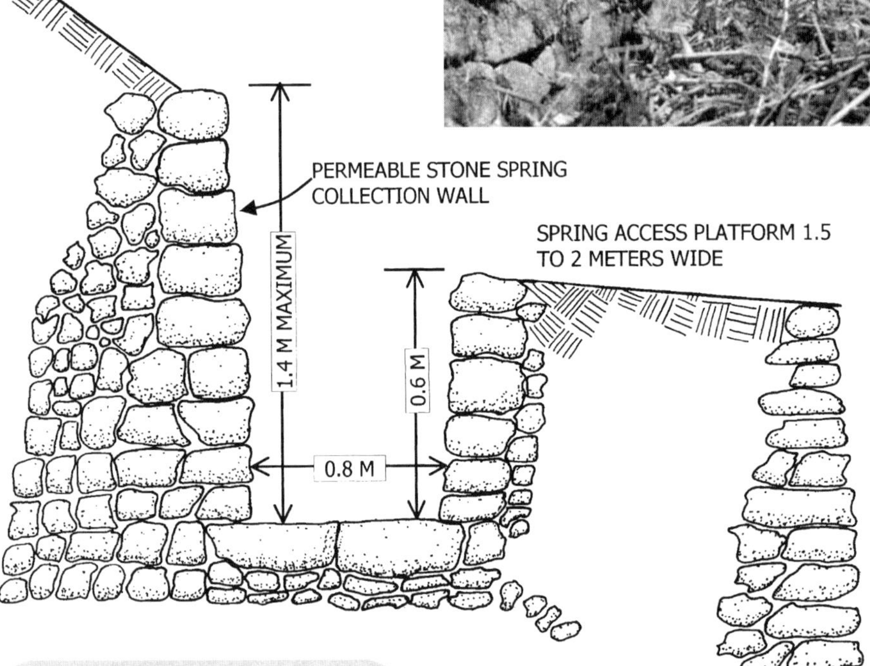

FIGURE 3-4▲
A cross-section of the Inca spring demonstrates the Inca engineers' high standard of care. Water flows in from the left to accumulate in the wide channel.

cross-section of the primary spring water collection system is illustrated in Figure 3-4.

Water of a secondary spring emerges 40 meters (130 feet) above the ancient Inca water supply canal and 80 meters (260 feet) west of the primary spring area. This water flows from the point of emergence to the Inca canal on the ancient canal terrace.

We searched for additional evidence of the Inca water collection system on the mountainside and conducted several field explorations into the dense forest area east of the primary spring. These explorations revealed no evidence that the canal extended farther east than the primary spring, even though the surface topography shows the existence of a significant gully beyond. As a result, we concluded that the water collection area works did not extend eastward from the primary spring. The Inca engineers apparently chose not to extend the system to the next defined surface drainage, although it would have been physically feasible to do so. About 10 meters (33 feet) east of the primary spring, we observed a wall of rounded cobbles that may have served as protection against earth slides or as a religious offering for the mountainside water source.

We took measurements of the primary spring yield and water quality at various times and compared them with average monthly precipitation at Machu Picchu. The winter months of May through August represent the dry season, while the summer months of November through March comprise the wet season. The relationship between wet and dry months and spring flow is provided in Figure 3-5. There is sig-

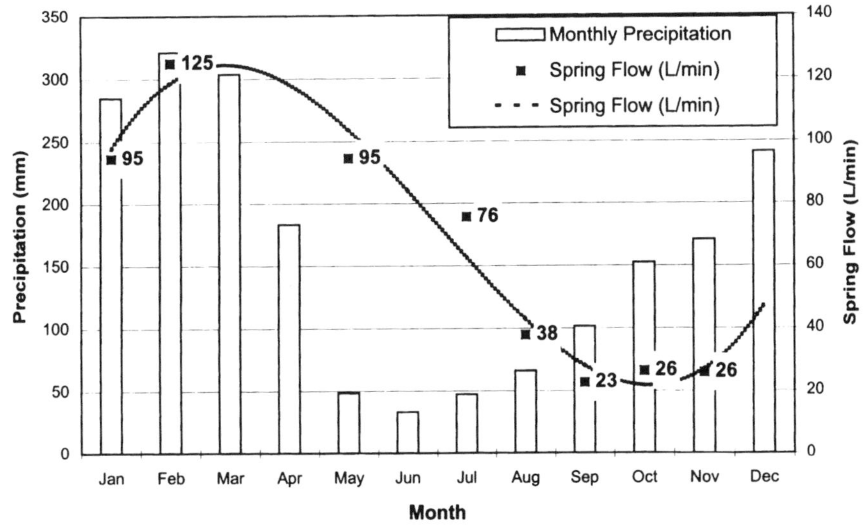

◄**FIGURE 3-5**
The water yield of the Inca spring was related to rainfall amount, but with a lag time of several months.

nificant variation in the flow from season to season. The spring yield is variable, ranging from a measured low of 23 liters (6 gallons) per minute to 125 liters (33 gallons) per minute. The variation indicates that the spring flow is derived from a relatively local hydrogeologic source that is influenced by seasonal variation in precipitation. This is consistent with the inflow–outflow evaluation and field surveys.

Water quality of the Inca spring is remarkably good (Table 3-1). Based on field measurements, conductivity generally ranged from 25 to 35 micro-Siemans per centimeter, pH ranged from 6.45 to 7.3, and temperature ranged from 14 to 16 °C (57 to 61 °F) as compared to the annual average air temperature of 15.6 °C (60 °F).

For many years, the original Inca spring was used to furnish water to the tourists who visit Machu Picchu. However, when the spring proved to be inadequate for the growing number of visitors, the government constructed a pipeline to bring in water from another source located several kilometers (miles) to the southeast.

Ancient Domestic Water Requirements

Based on the number of residential buildings, the ancient population of Machu Picchu is estimated to have been 300 permanent residents and up to 1,000 people when the Inca emperor was in residence (Hemming and Ranney 1982). Professor Rowe estimated about 40 normal dwelling rooms for the resident population, which tends to verify the 300-person estimate (Rowe 1997). The basic domestic water requirements of the residents could have been met by delivery of 10 liters (3 gallons) per minute to the fountains during the dry months. This is also the estimated flow required to routinely sustain an adequate jet of water to fill a ceramic water jug, or *aryballo*, at each of the 16 fountains (Figure 3-6).

FIGURE 3-6▲
The clay water jug, known as an aryballo, was used at the Machu Picchu fountains to collect water for carrying to the Inca homes. Water gathering was often a woman's job.

TABLE 3-1
Machu Picchu Domestic Spring Water Quality

	Units	*October 1994*	*February 1995*	*January 1996*
Inorganics				
Total Dissolved Solids	mg/L	40.00[1]	30.00	35.00[4]
Total Alkalinity	mg/L	11.10[1]	11.40	14.00
Total Kjeldahl Nitrogen	mg/L	<0.20[1]	<0.20	<0.20[3]
Ammonia-N	mg/L	<0.80[1]	—	<0.80[3]
Chloride	mg/L	<0.25[1]	—	0.87
Sulfur	mg/L	0.233[1]	—	4.42
Dissolved Metals				
Manganese	mg/L	<0.004	<0.004	<0.004 (0.0055)[2]
Copper	mg/L	<0.012	<0.003	<0.003
Zinc	mg/L	0.03 (0.04)[2]	<0.10	<0.10
Iron	mg/L	0.035	<0.04	<0.04
Aluminum	mg/L	<0.09	<0.09	<0.12
Sodium	mg/L	1.80	4.30 (2.10)[2]	1.80 (0.23)[2]
Potassium	mg/L	<0.41	<0.41	0.58
Calcium	mg/L	2.60	3.00	3.60 (0.26)[2]
Magnesium	mg/L	0.53	0.48	0.58
Total Metals				
Manganese	mg/L	0.009	<0.004	0.01 (0.0055)[2]
Copper	mg/L	<0.012	<0.003	<0.003
Zinc	mg/L	0.03 (0.04)[2]	<0.10	<0.10
Iron	mg/L	0.20	0.05	0.24
Aluminum	mg/L	0.33	<0.09	0.20
Sodium	mg/L	1.60	3.20 (2.10)[2]	3.00 (0.23)[2]
Potassium	mg/L	<0.41	<0.41	0.65
Calcium	mg/L	2.70	3.00	4.10 (0.26)[2]
Magnesium	mg/L	0.65	0.51	0.74
Radioactivity				
Gross Alpha	pCi/L	0.8 (+/− 1.2)	0.0 (+/− 1.1)	—
Gross Beta	pCi/L	3.2 (+/− 3.2)	0.0 (+/− 2.4)	—
Field Measurements				*July 1995*
Water Temperature	°C	14	14	16
Conductivity	μS/cm	30	25	35
pH		7.2	7.3	6.45

[1]Sample was filtered in lab with 0.45 micron filter to remove suspended material believed to be organic matter.

[2]This element was detected in the reagent blank. The blank value was not subtracted from the sample result. Blank value shown in parentheses.

[3]Analyses performed on an unpreserved sample.

[4]Sample received and analyzed outside holding time.

The Inca residents could have experienced insufficient water from the spring supply source during the dry months of the first and third decades of the period of occupation. Most likely, the Urubamba River, which lies 500 meters (1640 feet) below the community, served as a backup domestic water supply. The river has a reliable flow of high-quality water year-round. Original *aryballos* (Figure 3-7) were examined at the Cusco Regional Museum. The residents may have selected the appropriate size, made a trip down to the Urubamba River to fill

◄FIGURE 3-7
The aryballo could easily be carried uphill in the event water was gathered from the Urubamba River below. This Inca aryballo resides in the Cusco Regional Museum.

FIGURE 3-8▼
By looking down from the summit of Huayna Picchu to the east flank of Machu Picchu, one can see the route of the newly uncovered Inca Trail to the Urubamba River and the location of the 1998 fountain excavations by the authors. Shown is Terrace No. 4 built atop a steep cliff.

the jugs with river water, and climbed back up the 500 meters (1640 feet). The preferred route to the Urubamba River may have been via the wide and well-constructed Inca Trail that was discovered in 1998 by the authors and further described later. Here, the trail leads to a formal river landing which even today would be an ideal place for filling the aryballos. At various locations along the trail, additional springs and fountains flow even during dry periods. Thus, if the primary water supply were severely diminished by a reduced spring flow during the winter of a dry year, domestic water would have been available from these alternative water supplies, although they would have been much more labor-intensive sources.

The location and elevation of the primary spring water source are fixed where the spring exits the side of Machu Picchu Mountain. Therefore, the royal sector was likely planned and laid out after the spring was identified and the canal route and hydraulic slope were established so Fountain No. 1 and related buildings could be serviced by gravity flow.

Long-Forgotten Water Supplies

Remarkable discoveries were made between 1995 and 1999 far down the steeply sloping east face of the Machu Picchu ridge (Figure 3-8). Here, six fountain sites were uncovered while clearing paths through the rain forest (Figure 3-9). While the fountains were the focus of the exploration and excavation, an unexpected bonus was finding an adjacent Inca Trail common to three of the five fountains (Figure 3-10). This Inca Trail, long buried under thick rain forest, averaged 2 meters (7 feet) wide with granite stairways ranging to 3 meters (9 feet) wide. The trail was stabilized with retaining walls and

◀FIGURE 3-9
The water tunnel of Fountain No. 5 was explored and documented for the permanent records of the Instituto Nacional de Cultura. Here, Kenneth Wright is shown at the shattered opening of the tunnel. The damage was caused by a rockfall.

elaborate terraces to forestall slippage on the precipitous slopes hanging above near-vertical cliffs that dropped down to the Urubamba River valley below. This long-lost Inca Trail was likely the main trail from Machu Picchu to the Vilcabamba region lying further downstream in the Amazon rain forest and would have provided a convenient and efficient route for filling water jugs during a drought. This further proves that Machu Picchu was not abandoned because of a shortage of water supplies.

FIGURE 3-10▼
The long-lost primary Inca Trail from Machu Picchu down to the Urubamba River was discovered by the authors in 1998. After the rain forest was cleared, these granite stairways were found in excellent condition.

Groundwater was and still is the water supply of each of the lower east flank fountains. Having considerable drainage areas uphill of the fountains meant the Inca engineers did not have to work miracles for water. However, they most certainly had to carefully identify the specific dry-weather flow locations and tap into the seep areas to concentrate the flow at discrete points to support the fountain flows. As proof on the Inca's engineering skills, we were able to make the fountains flow again, even after 450 years of abandonment (Figure 3-11).

After the fountains were excavated and the site cleaned up, one of the Quechua Indians in the excavation crew, Florencio Almiro Dueñas, called the Wright Water Engineers team together for a traditional Inca thanksgiving prayer. The following prayer was spoken in the Quechua language, the ancient language of the Inca:

Today, having finished our excavations at Machu Picchu next to this water fountain, I call to the spirits of the Gods of Machu Picchu, Putucusi, Intipunku, and Mandor. Here, Pachamama—Pacha earth, beautiful mother, do not let the fountains go dry; every year water must flow forth so that we can drink. I am going to give you wine to drink.

"The Inca Were Good Engineers"

Because we took the time to examine details at Machu Picchu and think about the water supply challenges the Inca faced there, we can feel a sense of respect for these ancient civil engineers. They knew well the basic characteristics of the water supply prior to building the royal estate of Machu Picchu, as well as their probable water requirements.

Hiram Bingham was correct in 1913 when he told readers of the *National Geographic Magazine* that "the Inca were good engineers" (Bingham 1913). However, he could have also said the Inca were good civil engineers and hydrogeologists and knew how to put groundwater to good use.

FIGURE 3-11▲
Lower East Flank Fountain No. 4 was made to flow after it was excavated in 1998 and its approach channel was cleaned out. The niches and standard of care exercised on Fountain Nos. 3 and 4 indicated that they were ceremonial fountains.

Chapter Four

Hydraulic Engineering

The Inca were masters of technology transfer. The short, one-century life of the Inca Empire suggests they based their great hydraulic works on what they learned from the Andean and coastal people whom they overran during the 15th century, and their predecessors. Earlier South American civilizations built extensive water supply facilities and complex irrigation systems in Peru and Bolivia that surely inspired the Inca engineers.

While the Machu Picchu hydraulic works were not great in terms of size or liters per minute carried, they were "great" in terms of innovation, precision, success, and endurance—for even today they are able to function as they did when they were built.

Overview of Machu Picchu Hydraulic Works

Machu Picchu's urban focal point was the Temple of the Sun, a solar observatory and religious center with a series of 16 fountains (Figures 4-1 and 4-2). The aesthetic and functional layout and construction of the fountains make them a notable example of pre-Colombian civil engineering and planning. The fountains are in series except for the Sacred Fountain (No. 3) which now has an optional flow bypass to allow water delivery directly from Fountain No. 2 to Fountain No. 4. The bypass is likely a modern addition.

While the water supply for the fountains is derived from a natural spring, an effective spring collection works on the north side of the adjacent Machu Picchu Mountain (Figure 3-4) enhances the yield of this spring. A canal to carry domestic water by gravity from the Machu Picchu spring to the city center at Fountain No. 1, the uppermost fountain, is built on a series of stone-walled terraces. The locations of the water supply canal and the fountains are prescribed by the constraints of gravity flow of water from the primary spring (Wright et al. 1997c).

Water Supply Canal

The ancient Machu Picchu water supply canal tells us about the Inca ability to carry a suitable grade over a long distance, as well as their ability to build for the ages, even on steep, unstable slopes. Water was carried 749 meters (2457 feet) from the spring water source to the city center via a small domestic water canal formed with cut stones, as illustrated in Figure 4-3. For analysis, the supply canal is divided into

◄FIGURE 4-1
The fountains of Machu Picchu are in hydraulic series and each had an enclosure for privacy, a jet formed by an approach channel, a stone basin at the bottom and an orifice outlet that could be plugged. The adjacent long stairway provides access to each fountain.

four separate hydraulic reaches, shown in Figure 4-4. The domestic water supply canal immediately upstream of and in the agricultural sector is shown in Figures 4-5 and 4-6.

The bottom elevation, bottom width, top width, and depth of the canal were measured at 20-meter (66-feet) intervals throughout the length of the canal and representative dimensions were chosen for each reach. The typical slope of the canal varies from 2.5 percent to 4.8 percent. The canal cross-section area ranges from 125 to 168 square centimeters (20 to 26 square inches). The agricultural sector contains a 31-meter (102-feet) reach with a slope of 1.0 percent that limits the flow capacity of the canal in

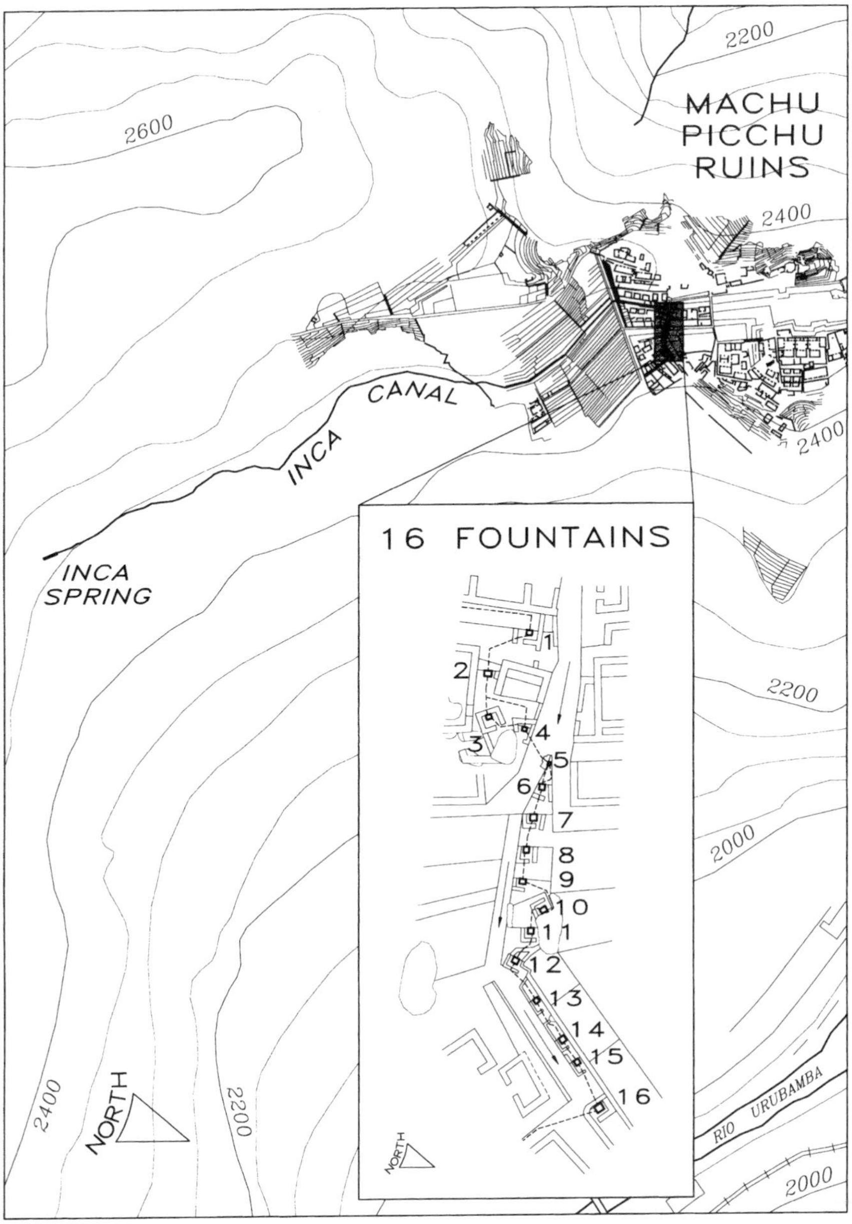

FIGURE 4-2►
The Stairway of Fountains parallels a series of 16 domestic water supply fountains that stretch over a distance of 55 meters (180 feet) with a vertical drop of 26 meters (85 feet). The fountains are similar, but the detail of each one is unique.

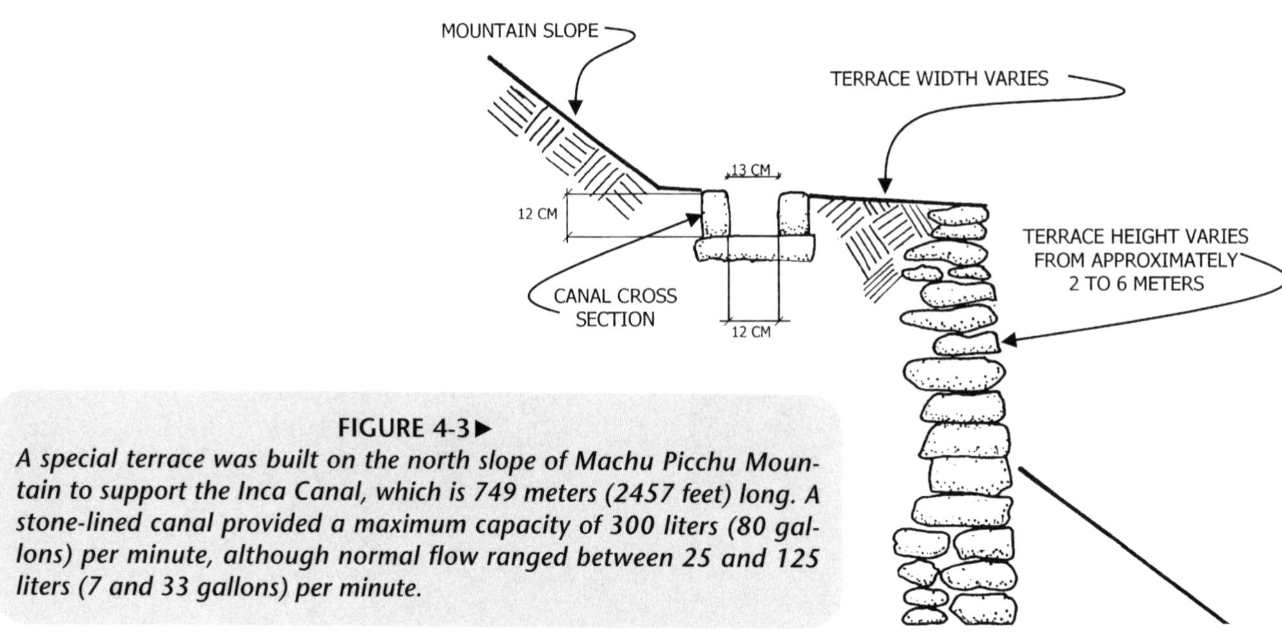

FIGURE 4-3►
A special terrace was built on the north slope of Machu Picchu Mountain to support the Inca Canal, which is 749 meters (2457 feet) long. A stone-lined canal provided a maximum capacity of 300 liters (80 gallons) per minute, although normal flow ranged between 25 and 125 liters (7 and 33 gallons) per minute.

FIGURE 4-4►
This is the plan and profile of Inca domestic water supply canal from the Primary Spring to Fountain No. 1.

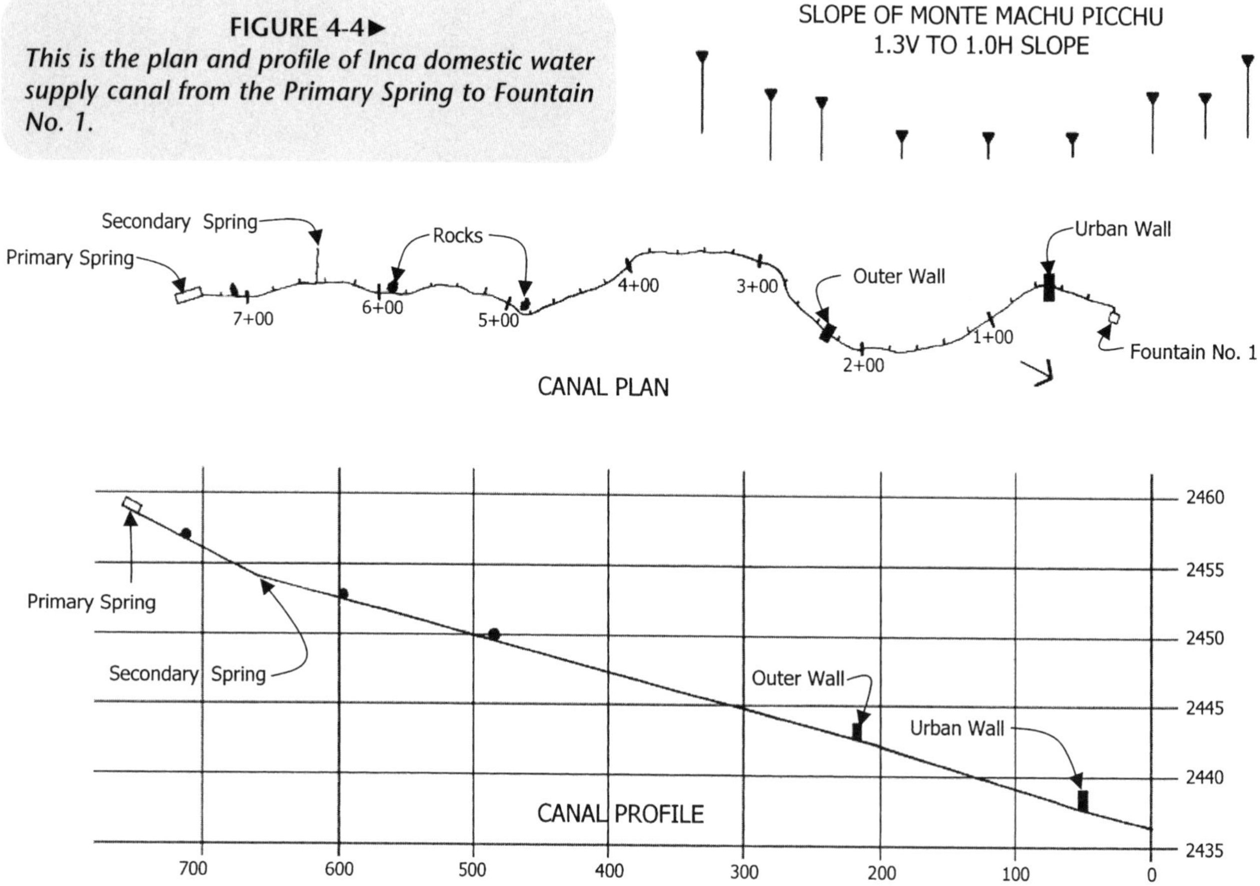

◄**FIGURE 4-5**
The route of the Inca Canal included support by a high terrace wall upstream of the agricultural sector announced by the two-story grain storehouse. The flow is moving away from the camera.

FIGURE 4-6▲
After leaving the forested hillside of Machu Picchu Mountain beyond, the Inca Canal courses through the agricultural sector in this stone-lined channel, flowing toward the camera.

this sector so excess water is spilled out onto the terraces. At two locations on the mountainside, field surveys of short reaches showed the canal invert to be at a flat grade. It is not known if this represents a settlement of the foundation over the centuries or if the canal failed due to landslides and was rebuilt with a slope sag. Bingham noted in 1912 that the canal failed due to landslides in several places (Bingham 1930). Because his photographs of the agricultural sector and urban sector showed the canal whole in 1912, the failure was on the mountainside.

The typical slopes and cross-sectional areas of the canal were reasonable considering the construction difficulties, steep terrain, variable spring yield throughout the year, and the need to avoid siltation and plugging of the canal with forest litter. Based on these characteristics, the authors judged the nominal design capacity of the canal to be approximately 300 liters (80 gallons) per minute. This flow is more than twice the maximum measured flow rate of the primary spring during the period of measurement and three times the capacity of the fountain system. Analyses have demonstrated that even at the canal "slope sags" on the mountainside, the canal capacity was adequate with a bank full capacity of 300 liters (80 gallons) per minute.

For hydraulic evaluation, we selected a nominal flow rate of 300 liters (80 gallons) per minute, keeping in mind that the Inca had no written language and would have designed the canal using empirical procedures. We calculated the hydraulic characteristics of the canal for each of the four reaches at the nominal design flow rate using Manning's equation:

$$Q = \frac{1}{n} A R^{2/3} S^{1/2}$$

where Q = flow capacity (m^3/sec); n = Manning's roughness coefficient; A = cross-sectional area (m^2); R = hydraulic radius (m); and S = energy slope (m/m).

TABLE 4-1
Hydraulic Characteristics of Inca Water Supply Canal at Nominal Flow of 300 liters (80 gallons) per Minute

Reach (1)	Length (m)[a] (2)	Typical Cross-Section: Bottom Width (cm) (3)	Top Width (cm) (4)	Depth (cm) (5)	Canal Slope (%) (6)	Flow Depth (%) (7)	Flow Area (cm^2) (8)	Velocity (m/sec) (9)	Froude Number (10)
Urban Sector	48	10	11	16	2.7	41	67	0.76	0.95
Agricultural Sector	153	12	12	12	2.9	68[b]	98[b]	0.52[b]	0.58[b]
Lower Mountain	461	12	13	10	2.5	54	67	0.74	1.0
Upper Mountain	87	12	14	10	4.8	44	53	0.96	1.5

[a] Total = 749 [b] Based on a limiting 31 meter reach at a slope of 1.0 percent.

A Manning's *n* of 0.02 was selected for the stone-lined canal based on evaluation of the canal roughness at 20-meter (66 feet) intervals. Table 4-1 summarizes the hydraulic characteristics for each reach of the domestic water supply canal at a nominal design flow rate of 300 liters (80 gallons) per minute. The hydraulic characteristics are indicative of a reasonable flow regime throughout the length of the canal.

Our hydraulic analyses of the Inca domestic water supply canal demonstrated that its builders included more than adequate capacity to carry the typical 25 to 150 liters (7 to 40 gallons) per minute yield of both the primary and secondary springs. The excess capacity provided a safety factor for most upper reach runoff into the canal, along with planned urban runoff from five drainage outlets in the urban sector. The size and slope of the canal meant that low flows in the 10 to 25 liters per minute (3 to 7 gallons per minute) range could be carried with reasonable efficiency.

The careful construction of the cut stone-lined canal was aimed at hydraulic and operational efficiency and control of seepage loss and, as a result, the maintenance requirements were relatively minimal. Seepage losses were likely not more than about 10 percent because of the well-fitted stone lining and the fact that the joints were probably sealed with clay. The canal is supported on terraces built on the steep slopes of Machu Picchu to resist sliding and settlement (Figure 4-3).

The infrastructure of Machu Picchu was carefully designed to maintain the purity of the domestic water supply. In general, the surface drainage layout directed agricultural and urban storm water discharges away from the open domestic water supply canal. However, the builders did construct one drainage outlet from the Temple of the Sun that discharges into the domestic water supply channel between Fountain No. 2 and Fountain No. 3. Also, four drainage outlets from the upper urban sector flow into the channel upstream of Fountain No. 1, but only after the drainage water flows over a short segment of vegetated land surface.

With regard to sanitation, Machu Picchu had no toilets, and no records describe how human wastes were managed; however, as with most ancient agriculturally oriented civilizations, human wastes would have been recycled to agricultural fields. Important nutrients would be

◄FIGURE 4-7
The well-engineered fountains of Machu Picchu have good flow characteristics for discharges between 10 and 100 liters (3 and 26 gallons) per minute. The sight and sound of the flowing water are a special treat for visitors. The approach channel at the top of each fountain was carefully shaped to cause the water flow to spring free to form a jet.

FIGURE 4-8▲
Adjacent to Fountain No. 3 (the Sacred Fountain), the Temple of the Sun was built with its Enigmatic Window overlooking the fountain. The Enigmatic Window of the Temple of the Sun is so named because of the numerous unexplained small holes in the adjacent stones, similar to holes in a door at the Temple of the Sun in Cusco. Some settling of the well-fitted stones of the temple is evident to the right of the window. The Guardhouse is in the background.

wasted without such practices. The Inca genius in corporate farming and the growing of food surpluses would have demanded knowledge of nutrient recycling. The standard of care the Inca used to protect the water supply canal from pollution indicates they learned the fundamental rules pertaining to "safe drinking water" from their ancestors.

Fountains

The operating fountains at Machu Picchu provide a special treat to even the most casual visitor (Figure 4-7). As in modern cities, the sight and sound of jetting water is an attraction—especially on a hot summer day. Machu Picchu's 16 fountains were laid out and designed to provide domestic water for the population, to enhance the urban environment, and as a manifestation of the power of the Inca ruler Pachacuti.

Fountain No. 1 was constructed adjacent to the doorway of Pachacuti's residence, providing him with the first opportunity to utilize the imported water supply (Figure 2-7). Fountain No. 3 is known as the Sacred Fountain because it is adjacent to a stone of adoration (*huaca*) and enigmatic window of the Temple of the Sun (Figure 4-8).

The Sacred Fountain has finely finished carved stone blocks and four niches for ceremonial objects. A specially

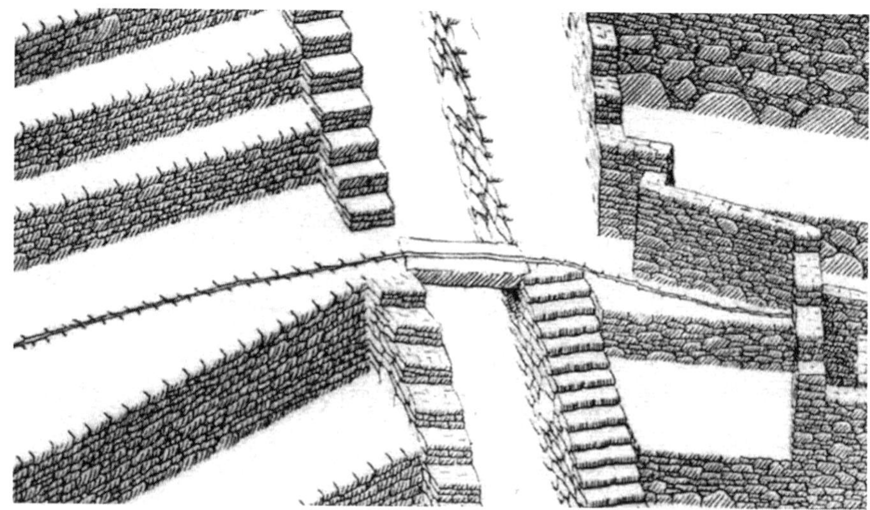

FIGURE 4-9▶
Hiram Bingham surveyed the newly discovered Machu Picchu in 1912 and described the Inca Canal as being on an aqueduct over the Main Drain, as shown above (Bingham 1930). Today, the canal is on fill across the drain.

carved rock adjoins the fountain on the east overlooking the eastern urban sector, the Urubamba River below, and the high mountains and sharp peaks in the distance. The carved rock near the jetting water would have provided an impressive backdrop for special ceremonies conducted by the Inca priests.

The water flows in series, from Fountain No. 4 to Fountain No. 16, until it is discharged to an underground stone conduit. From the stone conduit, water flows to the main Drain of Machu Picchu via a steeply sloping channel next to a long staircase.

The vertical drop between Fountain No. 1 and Fountain No. 16 is approximately 26 meters (85 feet). Except for Fountain No. 16, each of the fountains can be reached via common stairways and walkways. Fountain No. 16 is a private fountain accessible only from the Temple of the Condor.

We conducted detailed field instrument surveys, hydraulic flow tests, measurements of the fountain structures, and measurements of channel and outlet sizes. The small orifice outlet of 3.8 centimeters (1.5 inches) in diameter measured in the basin of Fountain No. 4 limited the maximum fountain system flow capacity to 100 liters (26 gallons) per minute. However, the individual fountains were designed to operate optimally with a flow of about 25 liters (7 gallons) per minute to fill the *aryballo*, the common Inca water jug. Field testing of the fountain hydraulic characteristics showed that the fountains would operate satisfactorily at flows as low as 10 liters (3 gallons) per minute. Hydraulic field testing showed that at flows of less than 10 liters (3 gallons) per minute, the water would not form a distinct jet, but would tend to flow down the stone facing. While the basin outlets could have been plugged to allow the basins to fill for easier water collection, the Inca would have used a plant leaf as shown in Figure 3-11. It is unlikely that the fountains were used for bathing and washing, for to do so would have been inconsistent with successive water uses from fountain to fountain. For instance, the emperor's residence has a special room for bathing with an outlet to the drainage system.

We investigated water control because the Inca residents would not have wanted excess water carried into the urban sector over and

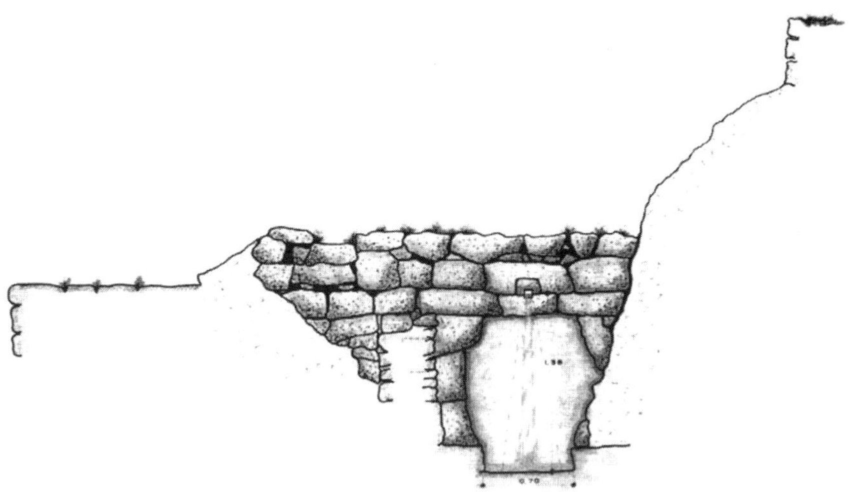

◀**FIGURE 4-10**
Machu Picchu Fountain No. 11 is shown with the entrance channel, jet, basin, and stonework. A granite outcrop supports the fountain enclosure on the right.

above the capacity of the fountain system. Two likely water control features existed. One, in the agricultural sector, consists of a 31-meter-long (102-foot-long) canal reach with a relatively flat 1.0 percent slope on its own narrow rock terrace. From this portion of the canal, excess water could easily be spilled over its right bank onto the agricultural terrace below. A second water control point existed at the main drain over which the canal was carried on a stone aqueduct spanning the moat just before the canal reached the urban wall (Bingham 1930) (Figure 4-9).

If excess water did reach the fountain system, the small outlet of 3.8 centimeters (1.5 inches) in Fountain No. 4 would have caused an overflow onto the granite stairway where the water would have been carried downhill for eventual disposal. This was verified in January 1996 when an intense early morning rainfall occurred. Mountainside and urban runoff entered the canal, spilled over the canal sides in several places, and overtaxed Fountain Nos. 1, 2, 3, and 4, causing the fountain jets to shoot beyond the stone barriers with high flows. However, the constraining orifice in Fountain No. 4 caused water to spill out of the fountain enclosure and onto the granite stairway where the water cascaded downward harmlessly until it dissipated onto the terraces.

While each fountain is unique in terms of its particular artistic character, all 16 fountains have the same general form, function, and layout. A carefully cut stone channel or conduit delivers water at or near the top of the fountain enclosure; a sharp-edged or lipped rectangular fountain spout creates a falling jet into a cut stone basin at the bottom of the enclosure (Figure 4-10). Each 3.8 to 5.0-centimeter-diameter (1.5 to 2.0-inch-diameter) circular-cut stone drain outlet carries the water to a channel leading to the next fountain. The walls of the enclosures are typically about 1.2 meters (4 feet) high to provide some degree of privacy within the fountain and to allow adequate height and floor space to conveniently fill the water jugs. Fountain No. 16, the private fountain for the Temple of the Condor area, is segregated from the public fountains via walls and stairway access. The enclosure walls for this private fountain are higher than the rest, rising about 1.6 meters (5.2 feet) from its basin level.

FIGURE 4-11▲
A secondary canal was planned for this terrace. Many stones in various stages of shaping lie where they were left when the workers departed in about A.D.1540.

Unfinished Branch Canal

Machu Picchu's abandonment in about A.D. 1540 was likely the result of the Spanish conquest and collapse of the central administration of the Inca Empire several years earlier. A few buildings were left partially completed, as was a new branch canal under construction on a terrace two levels below the main canal. Twenty-one stones in various states of carving and shaping (Figure 4-11) define the route of this canal on a special terrace.

The new canal would have been a branch off the main canal located near the entrance to the agricultural sector. The evident alignment indicated a planned canal upper reach of about 115 meters (377 feet) long, which would have traversed the agricultural sector and led to a proposed stone aqueduct crossing a large moat-like drainageway. The 22 additional channel-cut stones, including a bend, found beyond the drainageway, suggest a new series of fountains was planned (Figure 4-12).

The carved canal stones stretching out along the terrace demonstrate advanced stone-cutting skill. Each completed carved stone shows a relatively uniform trapezoidal hydraulic cross-section with a total area of about 21 square centimeters (3.3 square inches) formed by smoothly-carved sides and bottom. The cross-section is 3.3 centimeters (1.3 inches) deep with a bottom width of 5.0 centimeters (2 inches) and a top width of 7.5 centimeters (3 inches). The operating capacity of the unfinished branch canal would likely have been about 50 to 100 liters (13 to 26 gallons) per minute.

The fine carving of the channel cross-section in each of the scattered canal stones indicates that the Inca may have intended the new water system for ceremonial purposes. Certainly, the new canal would not have been for irrigation, as evidenced by its small size, fine workmanship, and lack of an irrigation water distribution network. To support this finding, our paleoclimatalogical and agricultural studies demonstrated that rainfall alone was adequate to support crop water needs.

FIGURE 4-12▲
The Inca engineers' planning for the branch canal included a stone-cut bend, which was abandoned and never used.

Hydraulics in a Nutshell

Machu Picchu is a pinnacle of the architectural and engineering works of the Inca civilization, which adopted public works technology from preceding civilizations and then carried the technology to new heights. The spring collection works, the main canal, and the 16 fountains of Machu Picchu represent the work of a civilization with sophisticated water-handling capabilities.

This chapter does not completely describe the hydraulics of Machu Picchu. For example, more could be said about our 1998 exploration and the clearing of the rain forest that uncovered wonderful additional fountains some 300 meters (980 feet) vertically down the steep east slope from Machu Picchu. Many potential new hydraulic structural and geotechnical discoveries likely are hidden in the forest awaiting the curious civil engineer/explorer.

Chapter Five

Drainage Infrastructure

The sophistication of a civilization can often be judged by its attention to the issue of drainage infrastructure, for it is the drainage system that suffers when the engineering standard of care is low. The Inca drainage infrastructure is one of the ways the Incas demonstrated that they built their cities for longevity.

The agricultural drainage and urban drainage systems at Machu Picchu were remarkable in their thoroughness and endurance. The standard of care the Inca civil engineers and workmen demonstrated was high enough to preserve Machu Picchu for modern research and tourism.

Public Works Achievement

The drainage infrastructure the Inca constructed at ancient Machu Picchu represents a significant public works achievement. The difficult site constraints associated with the nearly 2,000 millimeters (79 inches) per year of rainfall, steep slopes, landslides, and remoteness all posed drainage challenges that the Inca met successfully. The technical analysis of the Inca drainage works demonstrates that the drainage criteria used were reasonable and the implementation uncanny. It seems the Inca would have had the equivalent of an urban drainage manual similar to the ASCE Manual of Practice No. 77 (ASCE 1992) except that the Inca had no written language.

Agricultural Terraces

The agricultural terraces of the Inca royal retreat of Machu Picchu are visually dominating (Figure 5-1), complementing the magnificent structures of this mountain-top sanctuary both physically and aesthetically. The terraces also provided protection from excessive runoff and hillside erosion. This demonstrates that the Inca practiced sustainability some 500 years ago, which sets a good example for the world today in terms of soil stewardship (Wright and Loptien 1999).

The numerous ancient agricultural terraces total 4.9 hectares (12 acres). They are formed by stone retaining walls, contain thick topsoil, and are well-drained (Valencia and Gibaja 1992). The soil analyses showed that the sandy loam topsoil was thick; the deeper Strata II and III soils were more granular than the Strata I topsoil, which provided

FIGURE 5-1►
The visually dominating terraces of Machu Picchu were designed for beauty as well as function. Rainfall was infiltrated into the topsoil. The Guardhouse is at the upper right.

for higher permeability to enhance subsurface drainage. In deeper strata, the Inca workmen provided for excellent subsurface flow paths with loosely packed large stones and sometimes with stone chips from the stonecutting efforts.

We conducted field investigations to define the agricultural drainage. The canal that furnished water to the fountains in the urban area of Machu Picchu traverses the agricultural sector, but there are no turnouts to the terraces. We also performed field investigations to determine whether or not surface drainage water was reused for irrigation purposes. We found no evidence of reuse of surface drainage for irrigation within the Machu Picchu ruin. Neither was the discharge from the domestic water supply fountains directly reused for irrigation, but merely discharged to the main drain. However, we found evidence that subsurface drainage water was captured for several stone-formed fountains among the lower terraces downhill from Machu Picchu (Figure 5-2).

FIGURE 5-2▲
Recently discovered fountains on the lower east slope of Machu Picchu indicate that a steady flow of groundwater drainage was intercepted and put to use.

The technical study then proceeded as we evaluated modern climatological data and soils, estimated the likely ancient climate, and studied the agricultural drainage system of this royal estate (Wright et al. 1997). The surface drainage channels for the agricultural terraces are well laid-out with the contoured and terraced slopes suitable for shedding surface runoff to the adjacent drains. Our inspection of the terraces and the examination of photographs taken in 1912 by Hiram Bingham indicated little erosion from surface runoff, even after nearly four centuries of no maintenance and significant rainfall (Figure 5-3). The surface infiltration of rainfall to the subsoil drainage system was highly effective. The parallel surface drainage system for

FIGURE 5-3▶ *When Hiram Bingham cleared the terraces in 1912, he found little damage from erosion or wall failure. Bingham's 1912 tent camp is shown on the left in this* National Geographic Magazine *photograph by Bingham (1930).*

FIGURE 5-4▲ *Runoff from thatched roofs was sometimes captured with stone channels along the roof drip line, much like a rain gutter. Dr. Valencia Zegarra points to the former alignment of the overhead thatch. The Guardhouse with a thatched roof is shown at the top.*

the agricultural terraces, for the most part, provided redundancy and a drainage safety factor for intense rainstorm events.

Urban Sector

The urban sector of Machu Picchu covers 8.5 hectares (21 acres) and contains approximately 172 buildings, most of which were covered with thatched roofs. These residential and temple areas were laced with granite stairways and walkways, many of which also provided routes for drainage channels. The surface drainage outlets situated throughout the urban development in the retaining walls and building walls, when coupled with the drainage channels and subterranean caves, define the surface drainage network. The density of thatched-roof buildings (Figure 5-4) resulted in a generally high coefficient of runoff with a short time of concentration for the rainfall–runoff relationships.

Machu Picchu's drainage infrastructure and its special characteristics comprise the secret to its longevity. Archaeologists and scientists have long overlooked this fact, for without good drainage and foundation construction, not much would be left of the royal estate of Emperor Pachacuti. The buildings would have crumbled and many of the terraces would have collapsed due to the high rainfall, steep slopes, slide-prone soils, and settlement.

The Inca engineers gave high priority to Machu Picchu's surface and subsurface drainage during its design and construction. One may say the miracle of Machu Picchu is not only its beautiful buildings, but also the engineering features that lie unseen beneath the ground, where an estimated 60 percent of the Inca construction effort centered.

Studies of Machu Picchu identified 10 major components of its drainage system:

FIGURE 5-5▶
The centralized Main Drain separated the agricultural and urban Sectors. Putucusi is the rounded mountain in the background with the Urban Sector to the left.

1. A centralized main drain separating the agricultural sector from the urban sector (Figure 5-5).
2. Agricultural terrace surface drainage with reasonable longitudinal slopes, leading to formal surface channels integrated with terrace access stairways or to the main drain.
3. Subsurface agricultural drainage at depth typically consisting of larger stones overlain with a layer of gravel and above that a layer of somewhat sandy material (Figure 5-6).
4. Drainage management of the unused domestic water supply (Figure 5-7).
5. Positive surface drainage of urban grass or soils to drain the runoff from the many thatched roof structures and plaza areas. In some places, thatched roof drip channels exist.
6. Urban and agricultural drainage channels combined with stairways, walkways, or temple interiors (Figure 5-8).

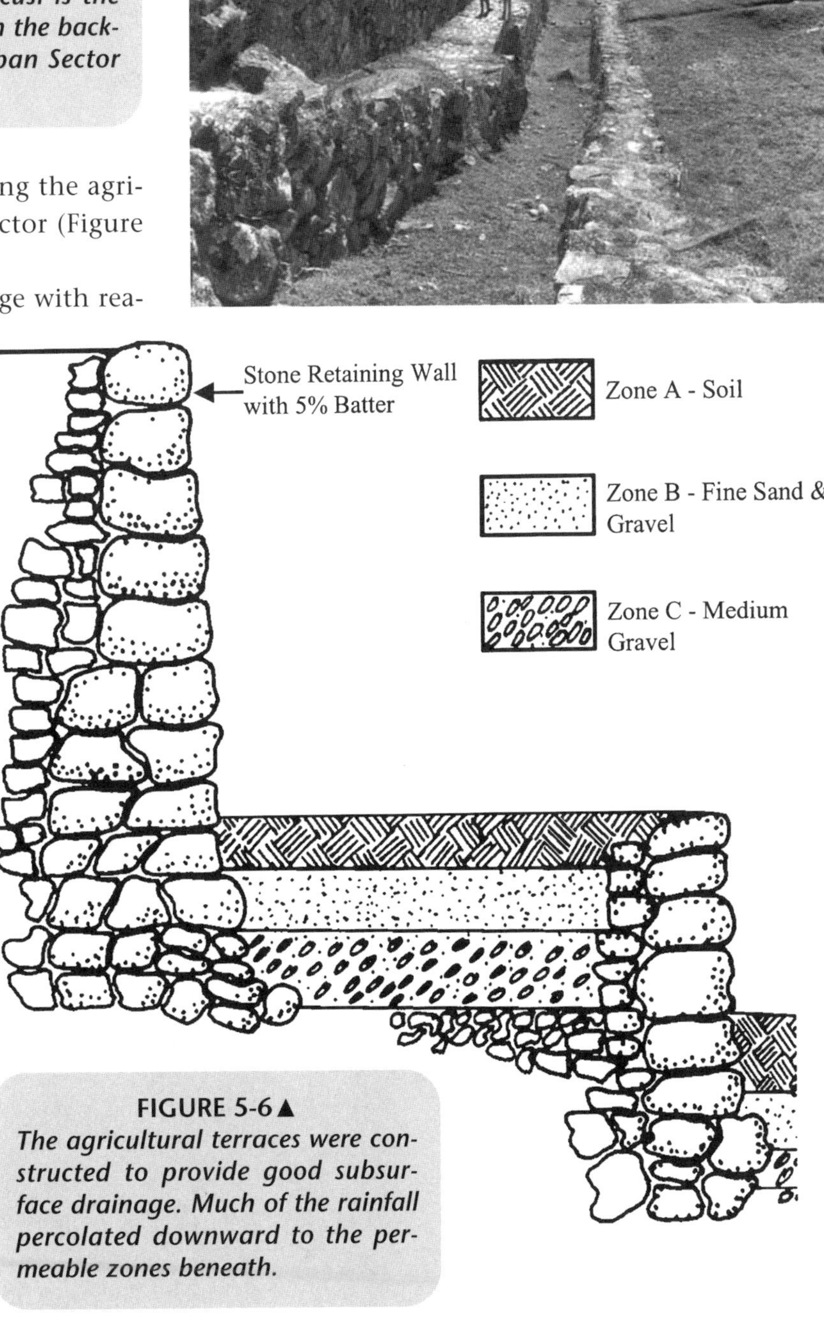

FIGURE 5-6▲
The agricultural terraces were constructed to provide good subsurface drainage. Much of the rainfall percolated downward to the permeable zones beneath.

FIGURE 5-7▶
Following Fountain No. 16, the unused domestic water supply is carried in an underground conduit for discharge to this staircase channel in which it is carried to waste in the Main Drain by gravity flow. Bingham described the channel in 1912 (Bingham 1930).

FIGURE 5-8▲
This granite stairway route provided a convenient path for safely carrying surface runoff from the top of the Intiwatana pyramid.

7. Deep subsurface strata under plazas of rock chips and stones (Figure 5-9) to allow the plaza to receive and infiltrate runoff from tributary areas.
8. A well-conceived and strategically placed urban area system of 129 drain outlets placed in the numerous stone retaining and building walls (Figure 5-10).
9. Subterranean caves with relatively free subsurface flow via natural underlying permeable deposits of decomposed granite and rocks.
10. Formalized systems for intercepting groundwater drainage on the lower east flank of the mountain ridge for serving ceremonial and utilitarian fountains (Figure 5-11).

◀FIGURE 5-9
Thousands of rock chips from the Machu Picchu stone cutting provided subsurface drainage, much like a modern French drain.

◄FIGURE 5-10
Urban storm drainage outlets at 129 locations throughout Machu Picchu ensured good removal of surface runoff. The drainage outlet on the left is in the Artisans Wall.

FIGURE 5-11▼
An archaeological excavation crew is shown at the lower east flank of Fountain No. 1 during 1998. The water supply for this lower lying water supply is partially supported by the flow in the Main Drain and groundwater interceptions from Machu Picchu.

Annual Water Budget

An annual water budget was used to estimate the likely amount of total runoff, both surface and subsurface, using the following approximate parameters:

- Average ancient period rainfall: 1,940 millimeters (76 inches) per year
- Agricultural vegetation evapotranspiration: 1,200 millimeters (47 inches) per year
- Composite unit urban area evaporation and evapotranspiration: 600 millimeters (24 inches) per year

The agricultural terraces yielded drainage water at the rate of 7,400 cubic meters/hectare/year (105,800 cubic feet/acre/year). The urban area drainage is estimated to have yielded 13,400 cubic meters/hectare/year (191,500 cubic feet/acre/year).

Based on examination of the topography, lack of erosion, type of soils, likely vegetation cover, and subsurface drainage potential, we estimated that about 90 percent of the annual water yield from the agricultural terraces occurred as subsurface flow and 10 percent as surface runoff. On the other hand, the area of urban building in Machu Picchu would have had approximately 60 percent surface flow and 40 percent subsurface outflow because of the impermeability of the thatched roofs and compacted soils. However, much of the

TABLE 5-1
Urban Surface Runoff Criteria for Typical Wall Drainage Outlets

Primary	*Magnitude*
Tributary area per drainage outlet[1]	200 m^2
Drainage outlet size, typical	10 cm by 13 cm
Drainage outlet capacity, maximum	650 l/min
Design rainfall intensity	200 mm/hour
Rational formula runoff "C"	0.8
Design flow per drainage outlet	500 l/min

[1]It was noted that the Temple of the Condor has only one drainage outlet for about 0.045 ha, more than double the area than that given in Table 6. However, the Temple of the Condor is underlain by subterranean caverns which drain most of the surface runoff.

surface flow is routed to the centralized plaza areas and to caverns for infiltration.

Based on the computed water balance and the field surveys of the Machu Picchu drainage system, along with an analysis of the agricultural terraces and urban area, we determined that the capacity and character of the subsurface drainage system was adequate. In terms of flow capacity and temporary detention storage, the groundwater component from relatively intense storms as well as runoff from an above-average wet year could be adequately managed without a resulting high water table.

Surface Runoff and Drainage Criteria

Using techniques based on trial and error and those transferred from one generation to another by word-of-mouth, the Inca builders were able to empirically size and design a remarkably effective surface drainage system. One cannot speculate as to how the drainage system was conceived or how the empirical criteria were developed; however, the long period of building by pre-Inca Andean empires such as the Wari and Tiwanaku would have given the Inca builders opportunity to judge and copy what would work.

We reformulated rough empirical Inca drainage design criteria by studying our map of Machu Picchu and analyzing urban drainage basins, by computing the capacity of the drainage outlets, and by documenting the spacing of the drainage outlets. We determined that a typical urban wall drainage outlet at Machu Picchu was based on criteria equivalent to those in Table 5-1. We do not assume that the Inca had formalized criteria; the parameters we developed for Table 5-1 represent their approximate empirical equivalents.

The Temple of the Condor has only one drainage outlet for about 0.045 hectare (0.1 acre), more than double the area of that given in Table 5-1. However, the Temple of the Condor is underlain by subterranean caverns that receive most of its surface runoff.

FIGURE 5-12►
An ancient buried stone wall was found under a Machu Picchu terrace, likely dating from the early construction period of Machu Picchu. Note the waste stone chips used for fill and subsurface drainage.

Plaza Subsurface Drainage

An Instituto Nacional de Cultura (INC) archaeologist conducted excavations for us at six locations to provide soil samples for laboratory testing. The archaeologist made three remarkable discoveries in Test Pit 6, which measured 2 meters by 2 meters (7 feet by 7 feet). They were:

1. An Ancient buried stone wall that represented either an early (A.D. 1450) change in construction plans or a temporary retaining wall for construction purposes (Figure 5-12).
2. A subsurface layer of loose rock and stone chips at depth for underground drainage. The rock chips found in the plaza represented a portion of the recycling of the thousands of cubic meters of waste chips from the stonecutters.
3. A gold bracelet that had been carefully placed between two guard rocks, apparently as an offering (Figure 5-13).

FIGURE 5-13▼
The first and only gold ever found at Machu Picchu was a fine gold bracelet that was discovered next to the ancient buried wall in Figure 5-12. The bracelet is on display at the Cusco Regional Museum.

The subsurface chipped rock layer was about 1 meter (3.3 feet) thick with an estimated coefficient of permeability of about 160 meters (525 feet) per day. The transmissivity of the chipped rock is then 160 square meters (1720 square feet) per day with a storativity of about 0.15.[1] As a result, the deep percolation from a major rainfall could be temporarily stored in the subsurface chipped rock layer where it would slowly drain at a modest rate to the downstream subsurface discharge point without causing a high groundwater table. Otherwise the plaza structure and its soils would have been unstable.

The plaza areas were also used to receive and dispose of storm drainage from adjacent urbanized tributary basins via

[1]In Meinzer units the parameters would represent a permeability of 4,000 gpd/ft^2 and a transmissivity of 13,000 gpd/ft.

FIGURE 5-14▲
Urban drainage outlets at Machu Picchu sometimes had vertical drains to avoid splashing and wall seepage.

◀FIGURE 5-15
Drainage outlets, such as this one on the Intiwatana pyramid, were carefully formed and placed to provide positive drainage.

infiltration and subsurface runoff. Again, as a safety factor, the Inca engineers also constructed surface drainage facilities for the plaza area.

Special Drainage Facilities

The descriptions of the general drainage features and criteria for Machu Picchu support Hiram Bingham's opinion of the Inca engineering capabilities. However, the investigation of specific drainage features demonstrates the special care the ancient designers took on a site-specific basis. Figures 5-14 and 5-15 illustrate details of Inca stone wall drainage outlets at Machu Picchu. Specific facilities are described later.

Emperor's Residence

Much like the entrance to a modern shopping center where good surface drainage is vital, the entrance to the emperor's residence in Machu Picchu was well drained. A small channel was constructed from west to east at the one and only entrance to the Royal Residence. The channel then passed through a retaining wall via a drainage outlet where the water would drop about 0.7 meter (2 feet) to another west-east flowing channel with a 2 percent grade. After 5 meters (16 feet), a drainage outlet and channel from the emperor's bath area joined the main channel (Figure 5-16). The channel then penetrated another exterior wall and continued to another drainage outlet that discharged to a channel on the inside of a main pathway.

FIGURE 5-16▲
The bath area of the Royal Residence had an individual drainage outlet that connected to the drainage channel running from left to right.

Because of its fine construction and stonework, the emperor's entrance drainage system was long thought to be an additional fountain somehow supplied with an underground conduit from Fountain No. 1. Even today, scientists and tour guides sometimes refer to it as a

◀ FIGURE 5-17
To avoid thatched roof runoff from percolating into the foundation zone, a drip channel was used to intercept the water. Here, at the Sacred Rock wayrona, the Inca thatched roof was much thicker than that of the roof thatch reconstructed by Dr. Valencia Zegarra.

fountain. The surface drainage at this location provides a good example of the special care taken to ensure the Emperor's doorway would be kept free of standing water.

Sacred Rock

The Sacred Rock is flanked by two, three-sided thatched-roofed buildings called *wayronas* (Valencia Zegarra 1977). Here, most of the drainage from the two thatched roofs drops onto the small plaza between the *wayronas* where peripheral surface drains carry the stormwater west to a low retaining wall with three wall drainage outlets. From there, the drainage discharges to the large plaza area separating the east and west urban sectors, thus avoiding surface water ponding.

A drip and drainage channel is carved into a large rock behind the southern *wayrona* (Figure 5-17). Water from the roof would drip onto the rock and be carried away from the *wayrona's* foundation.

FIGURE 5-18 ▲
This long, high-capacity drainage channel leading to the Main Drain was built in an area where the Inca engineers experienced an early earth slide that was then satisfactorily stabilized. The Inca Canal is shown on the right.

Agricultural Terrace Horizontal Drain

The Inca builders experienced ground slippage while the agricultural terrace area adjacent to the main drain was under construction and during occupation. The ground features were satisfactorily stabilized, but not before the Inca workmen took unusual steps to control surface water runoff in much the same way modern highway builders do today.

The builders constructed a large cross-section interceptor surface drain from south to north at the base of a long slope and immediately above the domestic water supply canal (Figure 5-18). This drain pro-

◄FIGURE 5-19
The Inca engineers used the Main Drain, here looking upstream, to collect runoff from the agricultural and urban areas for discharge into the rain forest below. The sharp stone on the right likely served as a survey marker.

tected the canal from potential overland flow pollution. The steep sloping interceptor drain, about 42 meters (140 feet) long, terminated at the main drain after passing through its south wall (Figure 5-19). Much of the uphill slope of this formerly unstable area is void of agricultural terraces, indicating that the builders of Machu Picchu had delayed constructing agriculture terraces in the area until they were certain that the steep slope would be stable. On the other hand, they may have decided never to build terraces there to avoid infiltration to the subsoils.

The Inca builders' measures were successful because the terraces were stabilized and the movement was stopped (Figure 5-20). A long granite stairway on the north side of the terraces, almost directly over the geologic fault, shows no signs of separation or damage, indicating that the stairway was built after the landslide was corrected.

◄FIGURE 5-20
The agricultural terraces in the lower foreground have 1 to 2 meter (3 to 7 feet) offsets while the adjacent granite stairway is without separation, indicating that an earth slide paralleling the Main Drain was stabilized during the construction process. Machu Picchu Mountain is in the background.

Chapter Six

Agriculture

Machu Picchu's agricultural terraces represent a civil engineering achievement in geotechnical, structural, and land planning technology, particularly when considering the longevity of the terraces. The Inca's ancient agricultural practices were so successful that people could be freed from subsistence farming to staff the military service and legions could be assigned to construct public works. Machu Picchu agriculture exemplifies Inca ingenuity and their successful use of terraced land.

Agricultural Terraces

The agricultural terraces of Machu Picchu provided rich planting soils on otherwise impossibly steep slopes. Not only did the terraces provide flat ground surface for food production, they protected against erosion and landslides common in the area and helped the Inca demonstrate their dominion over the land. The Inca civil engineers built the terrace walls so well that even after nearly four centuries, the walls were still intact when Hiram Bingham cleared Machu Picchu in 1912 (Figure 6-1). The terraces were constructed not only for agricultural purposes, but to create an amenity as they fit the hillsides and ridges like a well-tailored glove on a rough hand (Figures 6-2 and 6-3).

Questions frequently arise about whether the crops at Machu Picchu were irrigated, whether rainfall was adequate to support agriculture, and whether the agricultural terraces were capable of producing sufficient food for its population. During the investigation we determined that the rainfall of 1,940 millimeters (76 inches) per year was adequate for crops. We also found that the agricultural production of the terraces was not adequate to support the resident population and food would have been brought in from beyond the city walls. Field investigations determined that the agricultural terraces had no potential irrigation water sources; therefore, we then focused the technical study on the evaluation of modern climatological data, properties of the terrace soil, crop water needs, and potential nutrient production.

FIGURE 6-1▲
This 1912 view shows that the agricultural terraces at the time of their discovery were in good condition after nearly four centuries of abandonment. The photograph was furnished by National Geographic Magazine from its glass plate archives and represents two individual photographs from Bingham's (1930) composite.

Climatological Data

Machu Picchu's weather was monitored from 1964 to 1977. Subsequently researchers from Ohio State University took deep core samples from a Peruvian ice cap representing 1,500 years of accumulation. This provided a starting point for reconstructing the likely climate during the Inca time. The overlapping years between the Machu Picchu records and the annualized ice cap data was a civil engineer's dream.

The available climatological data for the modern period were selected for our analysis of agricultural potential at Machu Picchu. As will be shown later, the precipitation in 1964 through 1977, was similar to that during the time Machu Picchu was occupied. Tables 6-1 and 6-2 present the monthly precipitation and temperature data recorded at Machu Picchu from May 1964 through December 1977, based on Servicio Nacional de Metereología e Hidrología (SENAMMI), Cusco, Perú.

Machu Picchu lies about 13 degrees south of the equator and therefore has seasons opposite those in the United States. It has a dry winter season (May through August) and a rainy summer season (October through March). The temperature of Machu Picchu is mild and the area has no frost period. Seasonal temperature variations are considered modest. The precipitation and temperature characteristics are summarized in Table 6-3.

The ancient climate was estimated by using information from core samples from the Quelccaya ice cap located approximately 250 kilometers (155 miles) southeast of Machu Picchu (Figure 6-4). These data reflect the general historic regional climate affecting ancient Machu Picchu and were used in estimating its former climate. The Byrd Polar Research Center at Ohio State University conducted detailed and extensive ice cap and glacier studies and recovered two cores from the ice cap (Thompson et al. 1985). One core represented the years A.D. 470 to 1984 and the other represented the years A.D. 744 to 1984. These ice-coring data define wet and dry cycles extending from A.D. 470 to 1984 (Thompson et al. 1984). The long-term (A.D. 470 through 1984) average annual ice layer thickness for the two ice cores is about 1.4 meters (4.6 feet), estimated to represent 1,990 millimeters (78 inches) of annual precipitation. The 1964 through 1977 ice cap record

◄**FIGURE 6-2**
When viewed from the summit of Huayna Picchu, the agricultural terraces appear to be environmentally integrated into the natural topography so Machu Picchu is like a quilted blanket laid out over the ridge.

FIGURE 6-3▼
The work of the Inca engineers, as viewed from Uña Picchu, resulted in topographic relief that simulated the natural ridge top, yet created a structured urban environment.

TABLE 6-1
Monthly Precipitation at Machu Picchu (millimeters).

Year	*Jan.*	*Feb.*	*Mar.*	*Apr.*	*May*	*June*	*July*	*Aug.*	*Sep.*	*Oct.*	*Nov.*	*Dec.*	*Annual*
1964	—	—	—	—	48	24	23	54	71	58	108	216	—
1965	231	301	248	139	0	23	52	27	191	142	103	201	1,658
1966	242	272	148	58	32	2	23	30	37	171	225	248	1,489
1967	287	286	396	115	47	7	63	83	114	293	108	289	2,088
1968	350	399	353	107	31	30	132	81	65	197	313	210	2,268
1969	254	260	309	263	59	111	15	—	—	141	159	280	—
1970	—	329	288	268	108	28	79	34	108	160	118	283	—
1971	286	347	256	161	20	37	16	70	24	120	115	247	1,700
1972	381	227	344	187	63	2	32	94	92	118	262	263	2,064
1973	301	449	380	311	30	24	42	121	169	118	195	258	2,398
1974	258	356	214	258	20	27	65	126	102	178	79	142	1,823
1975	268	368	309	148	124	86	22	42	113	166	212	303	2,162
1976	276	260	399	—	—	—	—	—	—	—	—	—	—
1977	—	—	—	—	—	—	52	32	132	121	223	198	—
Average	285	321	304	183	49	33	47	66	102	153	171	241	1,955
# Years	11	12	12	11	12	12	13	12	12	13	13	13	
Min.	231	227	148	58	0	2	15	27	24	58	79	142	1,489
Max.	381	449	339	311	124	111	132	126	191	293	313	303	2,398

Source: SENAMMi (Servicio Nacional de Meteorología e Hidrología) Cusco, Perú.

shows an average ice layer thickness of 1.38 meters (4.5 feet), approximately the same as the long-term (NGDC, NOAA 1986). On the average, then, modern Peruvian precipitation is similar to ancient Peruvian precipitation. The decadal precipitation for the 90-year period of occupation of Machu Picchu, as represented by ice layer accumulations, is presented in Table 1-1; the 1964 through 1977 period ice accumulation is shown for comparison.

The first 50 years of occupation at Machu Picchu were drier than the last 40 years. A marked increase in ice layer thickness began in about 1500. This coincided with the well-documented Little Ice Age in the Northern Hemisphere (Thompson et al. 1986). The precipitation for the A.D. 1450 through 1500 period, based on annual ice pack accumulation, was approximately 1,830 millimeters (72 inches) or about 8 percent below the year A.D. 470 to 1984 long-term average of 1,990 millimeters (78 inches). On the other hand, the A.D. 1500 through 1540 period was 5 percent wetter than the long-term average with an equivalent annual precipitation of about 2,090 millimeters (82 inches). The annual variations would likely have been just as severe as those indicated by the modern data (Table 6-3).

The examination of the Thompson data from the Quelccaya ice cap was a treat in that ice etched records dated back to the time of the Vikings and not long after the fall of Rome. The climatic ice cap records meant that we could also begin to understand the climate changes that might have affected even earlier Andean empires such as of the Tiwanaku, the Wari, and Moche people.

TABLE 6-2

Monthly Maximum and Minimum Temperatures in Degrees Celsius (°C)

	Jan.		*Feb.*		*Mar.*		*Apr.*		*May*		*June*		*July*		*Aug.*		*Sep.*		*Oct.*		*Nov.*		*Dec.*	
Yr.	*Min*	*Max*	*Min*	*Max*	*Min*	*Max*	*Min*	*Max*	*Min*	*Max*	*Min*	*Max*	*Min*	*Max*	*Min*	*Max*	*Min*	*Max*	*Min*	*Max*	*Min*	*Max*	*Min*	*Max*
1964	—	—	—	—	—	—	—	—	—	20.5	9.5	20.9	9.8	21.1	9.7	21.1	10.4	20.7	10.9	21.6	10.3	21.3	10.3	19.7
1965	11.1	19.7	11.4	18.9	—	20.0	—	21.1	—	21.9	—	22.2	—	22.4	—	22.7	9.5	21.0	11.7	21.5	10.9	21.9	11.6	20.0
1966	12.0	20.5	12.3	20.4	11.4	21.8	11.3	22.7	10.2	21.9	8.5	23.1	8.1	22.4	8.8	22.5	10.7	23.4	11.0	21.8	11.9	21.0	11.7	19.7
1967	11.3	19.0	11.1	19.8	11.2	19.6	10.9	20.2	11.0	21.8	7.8	21.7	7.2	21.4	8.9	22.8	9.7	22.3	10.3	21.0	11.4	22.2	10.7	21.4
1968	10.5	19.3	10.9	—	10.1	—	9.9	—	8.1	—	8.3	—	7.8	—	9.1	—	9.6	—	10.7	—	10.5	—	11.0	—
1969	10.4	—	11.0	—	11.2	—	10.8	—	10.3	—	9.8	—	7.5	—	—	—	—	—	10.5	—	10.7	—	10.6	—
1970	—	—	10.2	—	9.8	—	10.2	—	8.9	—	9.0	—	6.8	—	7.5	22.7	8.6	21.5	9.5	21.5	9.9	22.0	9.7	19.7
1971	8.9	19.8	11.2	18.3	11.6	19.9	10.9	20.1	10.2	21.2	8.6	21.4	8.0	21.7	9.5	22.2	10.6	23.2	11.2	22.0	11.2	22.2	10.9	20.5
1972	10.6	18.3	10.8	19.9	10.8	19.8	11.2	20.8	10.2	21.7	8.5	22.6	8.7	22.6	9.0	22.0	10.2	21.7	11.9	21.6	11.8	20.3	12.5	20.4
1973	12.5	19.9	12.7	0.5	12.7	20.8	12.2	20.9	10.9	22.6	9.7	22.2	8.9	20.6	10.3	21.0	10.3	21.0	11.2	21.5	12.0	20.0	11.4	19.3
1974	11.4	18.6	11.0	18.0	11.6	20.6	11.3	20.3	9.6	21.9	9.3	21.7	8.9	20.7	8.3	20.4	10.4	21.6	10.9	20.9	11.7	21.8	11.5	20.3
1975	10.7	18.3	11.5	19.1	11.6	19.6	10.8	21.1	10.3	20.8	9.0	20.7	7.4	20.8	8.5	21.6	9.9	21.0	10.3	21.8	11.3	20.7	11.0	20.4
1976	12.5	19.8	10.8	19.2	11.1	20.1	—	—	—	—	—	—	—	—	—	—	—	—	—	—	—	—	—	—
1977	—	—	—	—	—	—	—	—	—	—	—	—	9.2	21.4	9.7	23.0	10.6	20.6	11.5	22.0	11.6	20.2	11.6	21.1
Average	11.1	19.3	11.2	19.3	11.2	20.2	11.0	20.9	10.0	21.6	8.9	21.8	8.2	21.5	9.0	22.0	10.0	21.6	10.9	21.6	11.2	21.2	11.1	20.2
# of Yrs.	11	10	12	9	11	9	10	8	10	9	11	9	12	10	11	11	12	11	13	11	13	11	13	11
Min.	8.9	18.3	10.2	18.0	9.8	19.6	9.9	20.1	8.1	20.5	7.8	20.7	6.8	20.6	7.5	20.4	8.6	20.6	9.5	20.9	9.9	20.0	9.7	19.3
Max.	12.5	20.5	12.7	20.5	12.7	21.8	12.2	22.7	11.0	22.6	9.8	23.1	9.8	22.6	10.3	23.0	10.7	23.4	11.9	22.0	12.0	22.2	12.5	21.4

Source: SENAMMi (Servicio Nacional de Meteorología e Hidrología) Cusco, Perú.

Hand-Placed Soil

While walking over the remarkably attractive and sturdy terraces, the engineer cannot help but admire the rich topsoil underfoot (Figure 6-5). How could such topsoil have stayed in place for so many hundreds of years? The answer, of course, is the standard of care the multitude of laborers used when they selected the soil. They hand-placed it and properly compacted it behind stonewalls laid out to provide just the right longitudinal and lateral slopes. But what kind of soil was it?

TABLE 6-3

Summary of Machu Picchu Precipitation and Temperatures, 1964–1977

Precipitation	*mm*	*Inches*
Average annual	1960	77.2
Maximum annual	2400	94.4
Minimum annual	1490	58.6
Average May–August	196	7.70
Average October–March	1475	58.1
Temperature	*°C*	*°F*
Average	15.6	60.1
Maximum monthly	23.4	74.1
Minimum monthly	6.8	44.2
Average May–August (maximum)	22.0	71.6
Average May–August (minimum)	8.2	46.8
Average October–March (maximum)	21.6	70.1
Average October–March (minimum)	10.9	51.6

We determined the characteristics of the soil in the agricultural terraces from 16 soil samples tested in 1996 at Colorado State University (CSU) (Wright Water Engineers 1996). We requested permission from Peruvian government representatives in February 1995 to take six core samples from the terraces; however, the request was denied by the INC. Instead, INC archaeologist, Señorita Elva Torres began the formal excavations of six terrace sites and collected more than 100 soil sam-

FIGURE 6-4▶
The layer cake-like stratification of the Quelccaya Ice Cap shows the annual ice accumulations. Layering of the ice core goes back to A.D. 480, providing clues to ancient climate.

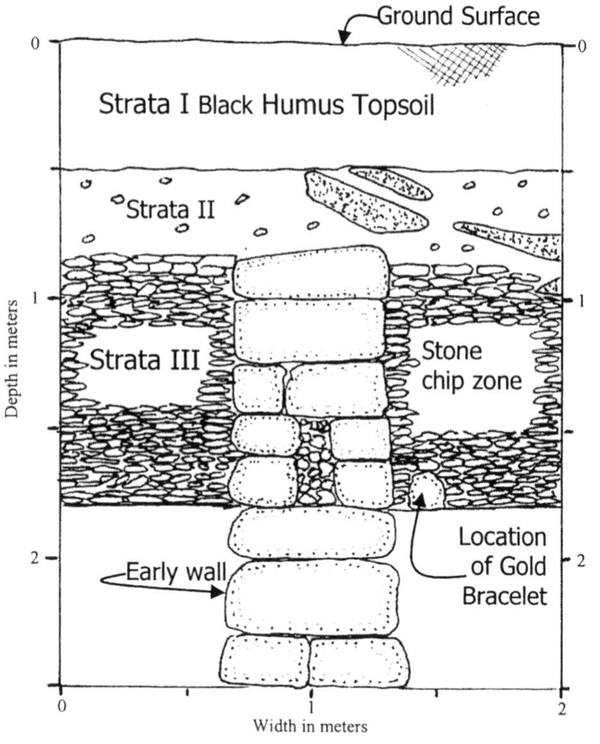

FIGURE 6-5▲
The soil profile below a flat terrace shows the topsoil placement and the more permeable sub-strata underlain with stone chips for good subsurface drainage. An ancient buried wall was built adjacent to this location, where a gold bracelet was discovered by Señorita Elva Torres.

ples from three strata at each test pit. In the sixth test pit, Señorita Torres discovered a gold Inca bracelet—the only gold ever found at Machu Picchu. The well-crafted gold bracelet currently resides in the Cusco Regional Museum (see Figure 5-13).

The authors received soil samples along with a government resolution authorizing their transfer to the United States. The CSU laboratory conducted a full array of agronomic tests on the ancient soils. A summary of the results is presented in Table 6-4. The soils were deficient in several nutrient and macronutrient constituents, probably as a result of four and a half centuries of in situ leaching from the substantial rainfall.

The excavations and laboratory test results show that the topsoil was typically 0.5 meters (1.5 feet) thick, and that even at greater than 1 meter (3 feet) deep, the subsoil of the deeper strata often contained topsoil characteristics. Fortunately, the gold bracelet discovery in Test Pit Number 6 justified additional archaeological excavations by the INC scientists during the fall of 1996. The subsequent and more extensive excavations there show a deep subdrainage system that represents the work of a well-organized and committed corporate agricultural society with knowledge of the importance of good drainage of soil for sustained agricultural production (Figure 5-20).

TABLE 6-4
Topsoil Gradation in Strata I Machu Picchu Agricultural Terraces

Particle Size	*Percent Retained*
>2 mm	15
1 mm	10
500 um	16
250 um	14
106 um	15
53 um	9
Constituent	*Percent*
Sand	66
Silt	17
Clay	17
Texture	Sandy Loam

Note: Bulk Density 1.3 g/cm^3. Weight Loss on Ignition 10%.

Crop Water Needs

After proving that the agricultural terraces were not irrigated, our next step was to determine the adequacy of precipitation to support the apparent agricultural production of the hundreds of terraces. Here, we turned to the ASCE Manual of Practice No. 70 (Jensen et al. 1990).

To judge agricultural production capability without relying on irrigation water for the Machu Picchu terraces, we needed to determine the crop water requirements so we could compare them with precipitation availability. In this case, we equated crop water requirements with evapotranspiration (ET) of the crops—that is, the water transpired by the crops during their planting and growth period and concurrent evaporation from the soil. Evapotranspiration is determined by climatological, agronomic, and solar radiation parameters and the use of computational techniques developed by the agronomy and engineering professions (Jensen et al. 1990). The basic climatological data presented in Tables 6-1 and 6-2 were used to determine the evapotranspiration of maize and potatoes at Machu Picchu. Evapotranspiration was computed for the following location and elevation of Machu Picchu: latitude 13° 09′ south, longitude 72° 32′ west, and 2,438 meters (8000 feet) elevation.

Solar Radiation

Solar radiation is important for determining crop water requirements (evapotranspiration) because it most directly relates to energy input resulting in evapotranspiration and crop growth. In computing solar radiation for Machu Picchu, the authors used data from two modern Peruvian towns (Figure 1-4) that have relatively similar elevations, latitudes, and longitudes to Machu Picchu and for which solar radiation estimates were available from CLIMWAT data files published by the

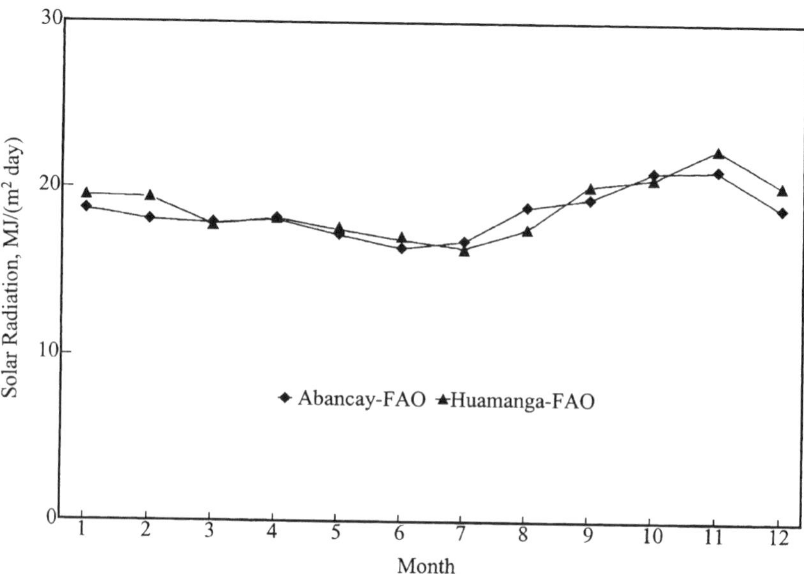

FIGURE 6-6▶ *Solar radiation, as measured at two Andean sites, was used to help determine crop consumptive use of water.*

Food and Agriculture Organization (FAO) of the United Nations (Smith 1993) (Figure 6-6). These sites are:

- Albancay
  Latitude: 13° 23′ south
  Longitude: 72° 32′ west
  Elevation: 2,398 meters (7868 feet)
- Huamanga
  Latitude: 13° 05′ south
  Longitude: 74° 08′ west
  Elevation: 2,761 meters (9058 feet)

The solar radiation data for the two Peruvian towns compared favorably in a check using our solar radiation estimates for the Machu Picchu sites. These estimates utilized extraterrestrial solar radiation (R_A), as calculated using standard equations (Knapp et al. 1990). This included a solar constant of 0.0820 megajoule per square meter per minute (MJ/(m^{2}min) and clear-day solar radiation of 0.75 R_A coupled with an elevation adjustment. Due to the favorable check comparison, we concluded that the average solar radiation data for the two Peruvian towns were suitable for transposition to Machu Picchu in determining energy availability for plant growth. We knew that in tropical climates, it is often better to use data from similar sites versus independent estimates for solar radiation.

Evapotranspiration

To estimate the ET of a specific crop at a particular location like Machu Picchu, estimates of the evapotranspiration of a reference crop are first obtained using the Penman-Montheith equation (Allen et al. 1989), and then coefficients that vary by crop growth stage are applied. Short grass and alfalfa were used as reference crops; the short grass was compared with the FAO estimate at Albancay and Huamanga (Table 6-5).

TABLE 6-5

Site Data for Machu Picchu, Grass Reference Characteristics, and Estimated Short Grass Reference Evapotranspiration using the Penman-Monteith Equation

Site Data for Machu Picchu
Latitude: –13.33 (south) or –0.23265242 radians
Elevation: 2438 m
Atmospheric pressure: 75.23 kPa

Assumed measurement heights:
Air temperature: 2.00 m
Wind speed (u2): 2.00 m

Reference Crop
Short Grass, height (hc) = 0.12 m
Momentum roughness parameter (zom) = 0.01476 m
Leaf-area-index (LAI) = 2.88
Surface resistance (rc) = 69.44 s/m

Aerodynamic parameters
Vapor and heat parameter (zov) = 0.1 zom = 0.00148 m
Displacement height (d) = 0.0800 m
Atmospheric resistance (ra) = 208 /u2, s/m

Input Data for Estimating Grass Reference Evapotranspiration (ETo)

	Month												
	1	*2*	*3*	*4*	*5*	*6*	*7*	*8*	*9*	*10*	*11*	*12*	*Average*
Solar radiation, (MJ/m² day)*													
Ra	40.47	39.70	37.37	33.56	29.70	27.60	28.33	31.47	35.34	38.40	39.96	40.44	35.19
Rso	32.33	31.71	29.85	26.80	23.72	22.04	22.63	25.14	28.23	30.67	31.92	32.30	28.11
Avg Rs	19.10	18.75	17.85	18.15	17.35	16.65	16.60	18.15	19.70	20.70	21.65	19.40	18.67
Mean temperature, °C													
T_{max}	19.3	19.3	20.2	20.9	21.6	21.8	21.6	22.0	21.6	21.6	21.2	20.2	20.9
T_{min}	11.1	11.2	11.2	11.0	10.0	8.9	8.2	9.0	10.0	10.9	11.2	11.1	10.3
T_{avg}	15.2	15.3	15.7	16.0	15.8	15.4	14.9	15.5	15.8	16.3	16.2	15.7	15.6
Mean precipitation, mm													
P	285	321	304	183	49	33	48	66	102	153	171	241	1955
Mean wind speed, m/s													
u2	2.56	2.82	2.82	2.78	2.89	2.8	3.1	2.5	2.4	2.55	2.55	2.45	2.69
Net solar radiation (Rns), and estimated net radiation (Rn), MJ/(m² day)													
Rns	14.71	14.44	13.74	13.98	13.36	12.82	12.78	13.98	15.17	15.94	16.67	14.94	14.38
Rn	12.08	12.40	12.21	12.64	11.99	11.41	11.38	12.64	13.83	14.33	14.48	12.21	12.63
Components of the Penman-Monteith equation, MJ/(m² day):													
Rad term	6.59	6.63	6.60	6.90	6.46	6.12	5.89	6.98	7.75	8.03	8.10	6.80	6.91
Aero term	2.19	2.33	2.62	2.87	3.47	3.74	4.17	3.46	3.00	2.92	2.73	2.37	2.99
Total	8.78	8.96	9.23	9.77	9.94	9.87	10.06	10.44	10.75	10.95	10.83	9.17	9.90
Estimated grass reference evapotranspiration, mm/day													
Eto	3.56	3.64	3.74	3.97	4.03	4.00	4.08	4.24	4.36	4.45	4.40	3.72	4.02
Average grass reference evapotranspiration for two similar sites†:													
	4.05	3.95	3.70	3.65	3.45	3.30	3.45	3.85	4.25	4.75	4.90	4.25	3.96
Estimated alfalfa reference evapotranspiration, (ETr), mm/day													
	4.27	4.36	4.49	4.76	4.84	4.80	4.90	5.09	5.24	5.34	5.27	4.46	4.82

* Ra = calculated extraterrestrial solar radiation, Rso = estimated cloudless day solar radiation, Rso = (0.75 + 2 Elev/10^5) Ra, and Avg Rs = average solar radiation. Average solar radiation obtained from FAO for two similar sites, Azancay and Huamanga.

† Average reference evapotranspiration from FAO for two similar sites, Azancay and Huamanga.

‡ ETo, mm/day × 1.2 = ETr, mm/day.

TABLE 6-6

Emergence and Harvest Dates Used to Estimate Evapotranspiration

Crop	*Emergence Date*	*Harvest Date*
Maize	1 November	Third week of April
Potato-1	1 September	End of January
Potato-2	Early January	End of May

Crop Evapotranspiration and Precipitation

The days required for emergence of maize and potato plants, leaf area development, growth at full crop canopy, and time to maturation were obtained by again using the ASCE Manual 70. Emergence and harvest data were chosen as shown in Table 6-6.

We estimated ET of maize and potato and daily crop coefficients. Figure 6-7 presents summaries of estimated crop ET for maize and potatoes, short grass reference ET (ET_r), and average monthly rainfall for Machu Picchu. The apparent short water supply for Potato-2 crop in May is not expected to have affected yield, since soil water accumulated during the preceding rainy months in the thick soil zone would be sufficient to enable the crops to mature adequately. The results indicate that the average Machu Picchu precipitation would have been adequate to provide the estimated average water requirements for maize and the two potato crops without the application of water for irrigation (Wright et al. 1997) (Figure 6-8).

FIGURE 6-7▼
This graph illustrates crop water requirements versus precipitation. The reference evapotranspiration (ET_r) is for short grass and is shown for comparison.

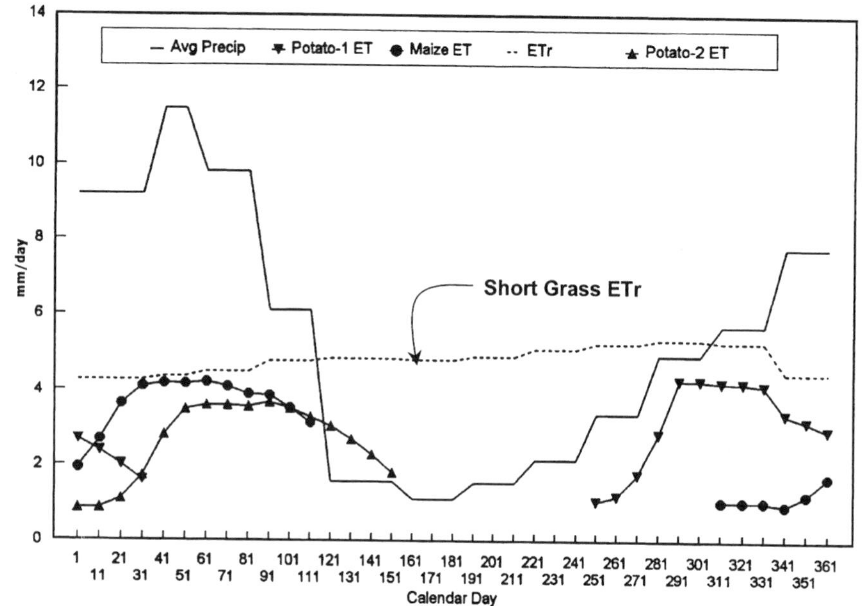

Adequacy of Nutrient Production

The analysis of the likely annual nutrient production at Machu Picchu was based on estimates of the ancient yield (pre-Green Revolution) of maize at 1,570 kg/ha and potatoes at 31,400 kg/ha with 2.5 ha of potatoes having two crops per year. The resulting annual crop production was computed:

FIGURE 6-8▶
The Inca potato cultivation methods are illustrated in this 16th-century sketch. Potatoes were a staple of the Inca people and are still a staple of their descendants.

Maize: 2.5 ha × 1,570 kg/ha = 3,925 kg/year

Potatoes: 2.5 ha × 31,400 kg/ha × 2 crops = 157,000 kg/year

Total = 160,925 kg/yr

The nutrient value of the annual crop production, using the U.S. Department of Agriculture estimate (Leverton 1959) of 16,300 kJ/kg for maize and 740 kJ/kg for potatoes was:

Maize: 3,925 kg/yr × 16,300 kJ/kg = 55.8 million kJ/year

Potatoes: 157,000 kg/yr × 740 kJ/kg = 116.2 million kJ/year

Total = 172 million kJ/year

Caloric requirements for an individual are estimated to be 2,000 kcal/day (8,400 kJ) or 3 100 000 kJ/year based on nutritional requirement studies (Adams 1975). Other investigators have estimated as little as 1,460 kcal/day as the minimum intake for Andean people (Kolata 1993); however, we concluded that an estimate of 1,460 kcal/day would be unreasonably low for the residents of a royal retreat in an empire whose success involved the provision of adequate food, and whose ability to produce surplus food was unparalleled at the time.

Thus, to determine how many people could be supplied with nutrients from all the terraced agricultural land within the Machu Picchu walls (Figure 6-9), the total of 172 million

FIGURE 6-9▼
A long-hidden agricultural terrace on the lower east flank of Machu Picchu is shown to the left in this photograph. The Urubamba River is below.

kJ/year of nutrients was divided by the 3.1 million kJ/year required per person, to arrive at 55 people.

The analysis that the agricultural land was capable of producing food for 55 people tells us that food had to be imported into Machu Picchu for the estimated 300 permanent residents. Even taking into account the slaughter of animals and the consumption of alpaca and llama meat, when the Inca retinue arrived from Cusco, bringing the population to approximately 1,000 people, even more food had to be brought into the royal estate,

FIGURE 6-10▲
Llamas help keep the grass at Machu Picchu trimmed.

Today, the terraced agricultural land produces an abundance of lush grass that is, in part, kept trimmed by the only modern residents of Machu Picchu, the llama (Figure 6-10).

Chapter Seven

Building Foundations and Stone Walls

The foundations and stone walls at Machu Picchu tell us a lot about the workmen and civil engineers of ancient Machu Picchu. The civil engineers selected and used a wide variety of wall styles and stonework quality to satisfy particular functions. Walls range from the finest—those of the Temple of the Sun and the tight-fitting polygonal stone of the Artisans Wall—to rough field stonework in the agricultural terraces.

Someday, when you wander through the buildings and passageways of Machu Picchu, keep your eyes open for the various types of stonework. The shape and character of the stone walls can provide you with a seemingly direct connection to the original workmen as they went about their tasks. You can even see stones that were left unfinished when Machu Picchu was abandoned in A.D. 1540.

Ancient Engineering Care

One of the reasons Machu Picchu endured through the centuries is because the Inca civil engineers used proven technology and a high standard of care in their building process (Figure 7-1). Much of their work is not visible until excavated (Figure 7-2). The unseen Machu Picchu exists in its subsurface drainage system and the extensive foundations that underlie the terrace walls and buildings (Wright 1996). Archeologists who have excavated at Machu Picchu conclude that 60 percent of its building effort went into the site preparation, drainage, and foundations.

After viewing the steep colluvial slopes of Machu Picchu, the civil engineer quickly realizes that the Inca relied upon at least a rudimentary knowledge of slope stability technology. Although analysis of the stability of a colluvial slope is perplexing, the success of slope stability at Machu Picchu demonstrates that the Inca engineers incorporated solutions common to today's engineers (ASCE 1969).

Foundations

The Inca technique for constructing wall foundations started with the careful placement of smaller rocks in the excavation bottom to create a firm bedding. The rocks became larger as the foundation rose nearer the groundline. A typical wall was 0.8 meter (2.6 feet) thick at the groundline, although the Inca civil engineers were not bound by such a rule. Some of Machu Picchu's structures demanded a more sturdy

◄FIGURE 7-1
Stable wall construction was an important Inca undertaking as shown in this 16th-century sketch. Note the standard of care indicated by the uniform offset vertical joints.

FIGURE 7-2►
This 2×2 meter (7×7 foot) excavation shows a hidden early wall that may have been a temporary retaining wall. The Inca used recycled stone chips for subsurface drainage conveyance and backfill under the terrace surface.

FIGURE 7-3▲
Archeologist Elva Torres points out a massive retaining wall that contained large foundation stones for stability.

thickness, such as the retaining wall east of the Royal Residence (Figure 7-3), which marked the edge of the western urban sector and the beginning of the central plaza. Here, huge stones were shaped, moved into position, and carefully fitted for stability.

In some instances the Inca engineers selected a "living" granite rock or huge in situ rocks for their foundation base, which was first shaped to provide a ledge or platform for founding. Examples include the Temple of the Sun (Figure 7-4), and the Temple of the Condor (Figure 7-5).

The *wayrona* building of the Sacred Rock represents use of in-place granite, wall foundation work, and extensive subsurface floor preparation as illustrated in Figure 7-6. In 1968, Dr. Alfredo Valencia Zegarra excavated and restored the building under the direction of Dr. Manuel Chavez Ballón. The site conditions were reported as follows (Valencia and Gibaja 1992):

Before construction, the terrain was very irregular—full of large and small rocks, giant boulders, and fragments of granite covered with vegetation characteristic of the humid, rainy climate of the "brow of the mountain." Then they built a retaining wall to the southeast to mark the boundaries and define the platform for the patio and of the enclosure. In this space, they first did a backfilling of large transportable rocks that covered some crevices and then smaller ones in order to complete the backfill. Then they began construction of the foundations, placing the base rocks over flat stones and rocks. Later, they continued the backfill with gravel so as to obtain uniform levels in the interior of the enclosure and patio. Over this layer, they placed fragmented and pulverized granite of a natural origin, a product resulting from the polishing and working of the stones.

As the construction advanced, the Inca required material for finishing the floor, which was yellowish earth, and which is characterized by its plasticity,

FIGURE 7-4▲
The curved-wall Temple of the Sun was built upon a huge granite outcrop after the top was flattened as a building surface.

FIGURE 7-5▲
Dramatic use of in situ rock for building purposes is exemplified at the Temple of the Condor where the rock and walls were made to appear as a large condor wing.

impermeability, and degree of compactibility. It was placed in the interior of the enclosure with a defined gradient toward the outside. They also put it in the patio, sloped towards the drainage channels; in this manner, they kept rainwater out of the interior of the enclosures, and prevented ponding in the patio.

The workmen constructed hundreds of ordinary terrace walls at Machu Picchu with various types of subsoil conditions ranging from decomposed granite and rocky soil to granite bedrock. Dramatic terrace walls are illustrated in Figure 7-7.

Stone Walls

In wall construction, the building stones were alternately placed in the wall lengthwise and then crosswise for stability. A header stone was placed at intervals to tie the front and back wall faces together. The stone was about 0.8 meter (2.6 feet) wide and extended from one face of the wall to the other.

Many of the stones have top or bottom indentations that helped fit them together in a nesting manner. Corner stones nearly always

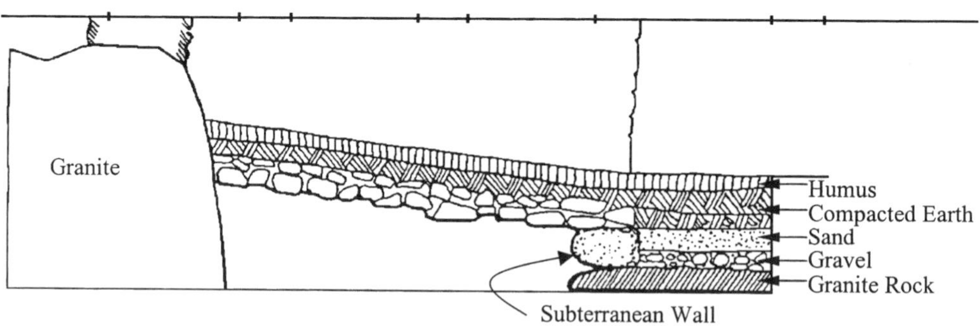

FIGURE 7-6▲
Dr. Valencia Zegarra's excavation of the **wayrona** *building at the Sacred Rock showed the foundation and floor preparation work performed at the site. The subterranean wall extended the full length of the building.*

have such indentations to provide additional stability. Walls were laid with a batter of 4 to 6 percent (Figure 7-8).

Mortar consisting of clay and earth mixed with small stones was used in many of the walls. Once workmen placed a course of stones over a layer of fresh mortar, they placed additional mortar in the center of the wall. Forgotten tools have often been found within the walls.

At Machu Picchu at least 18 different stone wall types and patterns range from the finest carving and shaping to the roughest type of permanent construction for use in agricultural terraces. There are even examples of rough, temporary construction stone walls used as an inclined plane for moving large blocks upward. Examples of each of the 18 stone wall types are shown in Figure 7-9.

▲FIGURE 7-7
Terrace walls on the west side of the Intiwatana pyramid demonstrate the Inca's care with slope stabilization.

FIGURE 7-8►
The batter of stone walls at an angle of about 4 to 6 percent helped create stability as shown here at the entrance to the Royal Residence.

FIGURE 7-9

Eighteen stone wall types are found at Machu Picchu, ranging from the finest masonry to rough field stonework. Each type is suited to a particular function or purpose. *(continues on next page)*

FIGURE 7-9
(continued)

◄**FIGURE 7-10**
This massive lintel beam in the Royal Residence contains a horizontal line that extends onto the adjacent walls. The purpose of the line remains one of the unsolved mysteries of Machu Picchu.

Special Building Stones

The many specially cut building stones add engineering interest to Machu Picchu. Lintel beams, recessed barholds for securing doorways, overhanging door rings, roof beam receptacles, and stairways are but a few examples, as illustrated in Figures 7-10 through 7-14. A special wall is shown in Figure 7-15.

Another special stone shown in Figure 7-16 is on the west side of the House of the Priests forming the north side of the entrance. The huge stone has many facets with a length of 3.1 meters (10 feet).

The west edge of the Sacred Plaza, where one can look down at the Urubamba River to the snow-capped mountain in the distance, includes specially carved and carefully shaped building stones that form a viewing platform (Figure 7-17). The wall faces the same direction and is similar to that of the Coricancha in Cusco.

An innovative use of shaped stones, geometric proportions, and orderly wall layout helped make Machu Picchu a World Heritage Site (Figures 7-18 and 7-19).

◄**FIGURE 7-12**
The close-up view of the ring and lintel beam in Figure 7-11 shows the care taken in stone shaping and fitting.

FIGURE 7-11▲
A ring above the only doorway to the Temple of the Sun, coupled with a bar hold on each side, provided a simple mechanism for a door closure.

FIGURE 7-13▶
A beam recess on the right at the Temple of Three Windows was carved into two huge granite stones.

FIGURE 7-14▼
A staircase carved out of a single block of granite at the Temple of the Sun provides a fine example of Inca stonework.

FIGURE 7-15▼
An hourglass-type wall was carefully fit between two large granite rocks at the Royal Tomb, or Mausoleum. Note the fitting of three cut stones to the rock at the upper right.

FIGURE 7-16▼
The house of the Priests at the Sacred Plaza contains a special stone with 32 angles.

FIGURE 7-17▲
The curved stones at the viewing platform on the west side of the Sacred Plaza are similar to those at the Coricancha in Cusco, and face the same direction.

FIGURE 7-18▲
The excellent stonework of the interior wall of the Temple of the Sun with geometrically spaced niches and wall pegs demonstrates a high standard of care in construction.

◄FIGURE 7-19
The geometric perfection provided by the Inca engineering is demonstrated by this view of doorways and niches within the Royal Residence.

Chapter Eight

Construction Methods

The Inca civil engineers did their work without the use of the wheel or iron. They had no draft animals for pulling huge rocks up inclined planes, yet they built Machu Picchu out of granite boulders, many weighing 10 to 15 metric tons (9 to 14 tons) and having lengths of 3 meters (9 feet).

Try to imagine the site conditions with which the civil engineers were faced in about A.D. 1450. Two mountains had a ridge between them and sheer drop-offs on both sides to the river 500 meters (1640 feet) below. They faced rock falls and jagged rock outcrops, nearly 2,000 millimeters (79 inches) of rainfall per year, and no level fields for growing crops. It was also a remote area some 80 kilometers (50 miles) from the capital, separated by steep passes and rugged mountain ranges.

In 90 years the Inca civil engineers changed this challenging site to the archeological jewel of the Andes, built so well that it endured for decades.

High Standard of Care

The Inca learned about long-lived infrastructure and building from their predecessors near Cusco at Pikillacta and Chokepukio. They also learned much about building technology from the Tiwanaku Empire centered near Lake Titicaca and how to avoid building in the floodplain, perhaps based upon the experiences of the Moche Kingdom on the north coast. From all of these earlier people the Inca learned the need to include solid foundations, good drainage, good agricultural soil, and a reliable water supply. As a result, their buildings have withstood the ravages of time and tropical rain forest tree roots (Figures 8-1 and 8-2).

The Inca must have planned Machu Picchu in stages. Their civil engineers would have first identified a suitable water supply, developed the water to "prove-up" its dependable yield, and then laid out a suitable canal right-of-way on a proper slope (Figure 8-3). Next, they would have ensured the location of a "first fountain" before being able to locate the Royal Residence next to the fountain. The engineers at Machu Picchu were remarkably successful with the water supply. They identified Fountain No. 1, set the location for the Royal Residence, laid out the series of 16 fountains perpendicular to the contours, and constructed the steep Stairway of the Fountains adjacent.

The main drain that separated the agricultural sector from the urban sector was the focal point of the drainage. It was designed to receive gravity drainage from both the north and south. The main drain was built on the line of a minor geologic fault extending up from the Urubamba River near the base of Putucusi Mountain.

◀ **FIGURE 8-1**
The exquisite stonework in the Artisans Wall, the long staircase, and the terrace walls to the left withstood the ravages of time and tree roots to endure into the 20th century.

The Inca civil engineers did not have perfect vision; they occasionally experienced difficulty going from plans to final construction. For example, a troublesome landslide along the northern perimeter of the lower agricultural terraces caused terrace walls to move downhill 1 to 2 meters (3 to 7 feet). Here, the engineers took steps to control the slide even after work on the terraces was well along. They carefully intercepted the surface drainage (Figure 8-4), refrained from constructing terraces uphill of the slide area to minimize water infiltration, and repaired the damaged terrace walls using offset curves (Figure 8-5). Finally, once the slide had been stabilized, the access stairway paralleling the northern edge of the terraces and adjacent to the main drain drainage channel was completed. Even today the stairway shows no evidence of the slide.

FIGURE 8-2 ▲
After Hiram Bingham discovered Machu Picchu he announced to the world in the 1913 National Geographic Magazine *that this long wall of the Temple of the Sun was "the most beautiful wall in America." It lasted through the ages because of the Inca's high standard of building.*

We also know that the Inca civil engineers sometimes changed their plans after construction had started as evidenced by an unused foundation wall discovered in 1969 by Valencia and the small abandoned subterranean area found in 1978 by Ramos in Conjunto 2 (Valencia and Gibaja 1992).

FIGURE 8-4▲
The Inca constructed this discharge orifice from the upstream interceptor drainage channel (Figure 5-18) to withstand hydraulic forces and erosion. Here, the authors are measuring its cross-section.

FIGURE 8-3▲
The uniform slope of the Inca Canal safely carried domestic water into Machu Picchu over a wide range of flows. The terraced right-of-way provided room for maintenance workers.

Rock Quarry

The Inca used numerous quarries to furnish the many thousands of cubic meters (tons) of stone; however, the principal remaining quarry is known as "Caos Granítico" located south of the Temple of Three Windows (Figure 8-6). The quarry contains blocks of granite of all sizes derived from ancient landslides and rockfalls.

The basic building stone is local igneous intrusive granite made up of quartz, feldspar, and mica. In several locations the Inca sparingly used a green talc stone, a product of the Machu Picchu fault, and pink granite imported from Ollantaytambo, some 50 kilometers (30 miles) upstream.

FIGURE 8-5►
The offset in the terrace walls of 1 to 2 meters (3 to 7 feet) supports other evidence that the Inca engineers experienced an early earth slide at a minor geologic fault. The slide was stabilized.

◄FIGURE 8-6
The quarry south of the Sacred Plaza provided granite building stones for Machu Picchu. The peak in the distance is Machu Picchu Mountain.

FIGURE 8-7▲
Natural fracture joints in the local granite rock provided opportunities for the Inca builders to take advantage of nature's work.

Quarrying of stone utilized the natural fracture patterns of the rock (Figure 8-7) along with the use of expanding wet wood wedges and fire for heat fracturing.

Transporting and Lifting Rocks

Architects Jean-Pierre Protzen (1993) and Vince Lee's (1999) careful analysis at ancient quarries and stone transporting routes provided insight into the transportation of large stones. In addition, our observations of how modern Quechua Indians—the local descendants of the people of the Inca Empire—move stones while participating in archeological excavations show their wise use of wooden poles, both as levers and as inclined planes.

The Inca workmen used slides to move large rocks downhill and inclined planes for transporting rocks uphill. To raise a large stone in elevation at a wall, they may have employed a "teetering" process, using levers and placing timbers and rocks underneath. Transporting large rocks required manpower and perhaps ropes, as illustrated in Figure 8-8. On the other hand, Lee (1999) surmises that log runners and log sleds were more likely used with the movement resulting from the application of multiple levers being applied in unison (Figure 8-9).

We observed a single Quechua Indian raising and moving smaller stones, from 100 to 150 kilograms (220 to 330 pounds), with nothing more than a wooden pole serving as a lever and alternately as an inclined plane.

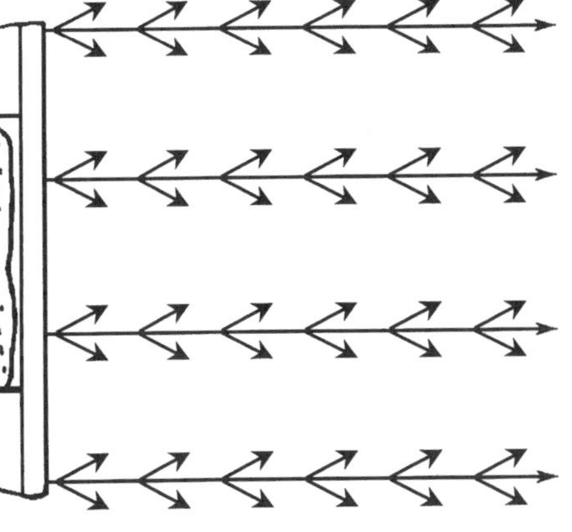

FIGURE 8-8▲
One theory is that a large crew of men and women dragged large stones by pulling them with ropes. Rounded cobbles were used under the large stones to reduce friction (Figure 8-14).

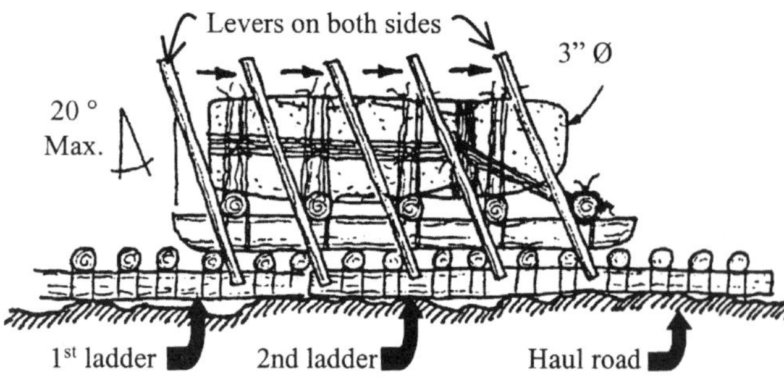

FIGURE 8-9▲
A more recent theory for large stone hauling involves the use of multiple levers. Architect Vince Lee from Wyoming field-tested the method shown here in Longmont, Colorado, in November 1998. He moved a large stone with a small group of volunteers.

Using Wood and Vegetation

The Machu Picchu area's varied flora consists of arboreal, conifer, latifolds, and alder species (*Alnus jorullensis*) that the builders used for girders and doorways. At lower elevations are tree ferns (*Cyathea and Alsophilla*) and reed grass (*Phragmites communis*), which the Inca used for thatched roofs. A variety of lianes were used for ropes to join the various beams/girders. The bronze axes that Bingham discovered at Machu Picchu in 1912 were likely used to cut the trees.

The workmen used wood in buildings, primarily for roof structures, doorways, stairways, second floors, and lofts. In the case of the roofs, they used thick trunks cut for the roof supports. They worked the wood until they had straight, square-cut girders or rafters. They joined the wood using ropes made of lianes or camelidae fiber.

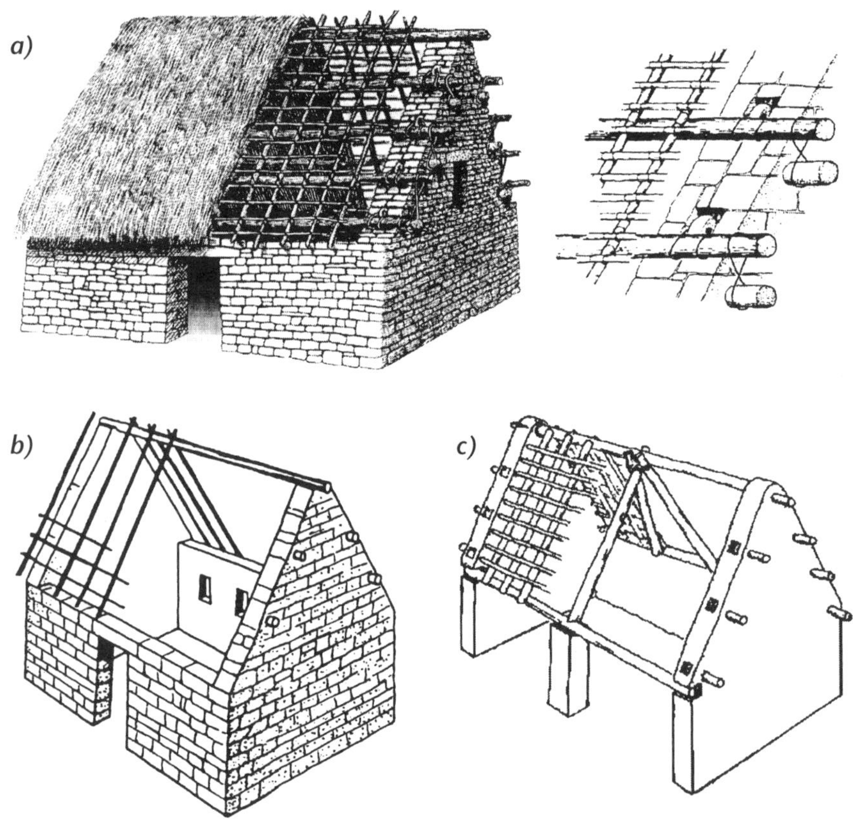

FIGURE 8-10▲
The enigma of how the Inca civil engineers designed their roofs has challenged modern structural engineers for decades. The three theories range in time from 1912 to 1988. Modern engineers have a fertile field of research in Inca roof technology. a) Bingham's 1912 analysis of attaching roofs using eye-anchors and roof pegs. b) Gasparini and Mangolies' 1977 analysis of the roof structure. c) Vincent Lee's 1988 analysis uses eye-anchors and roof pegs to hold down thatch roofs.

Roof Structures

Just how the Inca engineers designed their roof structures is still uncertain even though many clues exist in the form of pegs, rings, notches, gables, and stone supports. One would think that with so many clues available, the engineers, scientists, and architects would have figured this out. But this is one of the questions still under study. Students of Inca structural engineering have proposed three theories, as illustrated in Figure 8-10.

Canal Stones

Fieldstones shaped on one side formed the sides of the rectangular domestic water supply canal. Flat stones creating a tight joint that could be sealed with clay to reduce water seepage often formed the canal bottom (Figure 8-11). Careful examination of the canal reveals occasional flat, green talc stone forming the canal bottom. The use of this unique stone type is just another yet-unresolved mystery of Machu Picchu.

FIGURE 8-11▲
The canal stones needed only one flat side to form the channel of the Inca Canal. Joints could be sealed with clay to reduce seepage. Here, an Inca canal maintenance worker would have fallen some 12 meters (40 feet) by slipping off the narrow terrace.

Floors and Plaster

Limed clay plaster usually covered the walls of the rooms. The clay is a residual of the disintegration of igneous rock. The plaster covered extensive surfaces of the interior floors and patios. Plaster was sometimes used on rough exterior walls; however, walls with fitted joints were not plastered. The agricultural terraces were not finished or plastered.

The floors were placed over layers of sand, gravel, and large rocks. The thickness of the finished floors varied from 20 to 30 centimeters (8 to 12 inches).

Archeologist Luis Astete analyzed a plastered wall in the interior of the enclosure of Conjunto 1 of the urban sector. The wall had three layers with the following characteristics: the first layer adhered to the stone wall formed of a clay containing fine stones; the second layer, a natural sienna color, was finer than the first layer; and the third layer, an ochre yellow color, had been fired. The plaster was 2.5 centimeters (1 inch) thick (Valencia and Gibaja 1992).

The plaster likely covered the walls and floors uniformly—perhaps to the top of covered structures and over any wooden lintels to make them impermeable to the rainwater and the humidity, as well as for protection against insects.

FIGURE 8-12▲
The Principal Temple included huge, well-fitted granite blocks. The temple was not completed due to foundation problems and settlement of the east wall. The large rock in the center served as an altar-like platform.

Abandoned Construction Work

The civil engineer who visits Machu Picchu and wants to view construction in progress as of A.D 1540, should inspect several unfinished projects.

Principal Temple

The side walls of the Principal Temple were not completed (Figure 8-12). The Principal Temple would have had a sloping roof to the rear wall, creating a large *wayrona*. Work was likely put on hold due to the settlement of the massive east wall—the foundation being no match for the extremely heavy weight of the wall. Settlement during Inca time is illustrated in Figure 8-13. The outside northeast corner of the temple still stands as if waiting for the stonemasons to return to complete the shaping of the in-place corner stones.

Temple of Three Windows

A large stone at the Temple of Three Windows was being moved when work stopped at Machu Picchu (Figure 8-14).

◄FIGURE 8-13
The 0.3 meter (1 foot) settlement of the east wall of the Principal Temple is evidenced in the north wall where the stones have been pulled downward and out. Bingham recorded this settlement damage (Figure 2-17) in 1912.

◄FIGURE 8-14
At the Temple of Three Windows, the Inca workmen were moving the large stone in the right foreground when they abandoned work in about A.D. 1540. The small stones underneath reduced dragging friction. Behind photographer Ruth Wright is a special roof beam pedestal to support a beam extending to the recess at the right.

Unfinished Wall near the Granite Spires

This work in progress still has the rudimentary inclined plane in place and, upper left, one large stone is still tipped up at 45 degrees waiting to be finished and set in place (Figure 8-15).

Branch Canal

When Machu Picchu was abandoned, the stone workers were busy building a branch canal. This abandoned canal would have begun near the upper grain storehouse in the agricultural sector extending northerly to the main drain and beyond. Twenty canal stones are lying on the terrace south of the main drain; some stones are completed and some are in various stages of shaping. The stones are spread out over 40 meters (130 feet) of the proposed terrace route lying just below the primary completed canal and attest to the terrace route as it crosses the agricultural sector. At the north end an aqueduct stone would have crossed the main drain and then continued downhill where there likely would have been another series of fountains. Figure 8-16 shows

FIGURE 8-15►
This stone wall was under construction in about 1540 A.D. The large stone above and to the left of the survey rod is tilted back at 45 degrees to allow workmen to shape its bottom to fit the two stones underneath. Note the uniform corner construction to the right.

FIGURE 8-16►
This aqueduct stone has a near perfect shaped trapezoidal channel carved from end to end. The channel cross-section matches the dozens of other channel stones strewn along the right-of-way of the abandoned branch canal.

FIGURE 8-17▲
This partially completed channel stone illustrates the stonecutters' technique. The supervising cutter would chip out the two ends of the channel, which would then be connected by a less-skilled workman.

FIGURE 8-18▲
Even Inca engineers can have a change in plans. Here, at the Intiwatana, a doorway was converted to a window.

FIGURE 8-19▲
This 16th-century sketch of an Inca bridge shows a structure similar to swinging bridges still used by modern Quechua Indians. Many abutments still exist from the Inca Empire.

the aqueduct crossing stone with its finely finished canal section complete from end to end. Forty-four stones are located on both sides of the main drain, with a channel fully or partially cut for use in the abandoned canal.

Measurement of the scattered unfinished canal stones shows that the channel dimensions and shape were carefully controlled by the Inca civil engineers. The stones have a uniform trapezoidal shaped channel 7.5 centimeters (3 inches) across the top and are 3.3 centimeters (1.3 inches) deep, as shown in the partially completed stone in Figure 8-17.

Change Orders

The wall of a small temple on top of the Intiwatana provides a good example of a "change order." Here the wall was originally constructed with a fine trapezoidal door. Later, the door was changed to a window (Figure 8-18).

Inca Bridges

The Inca constructed bridges across the Urubamba River, perhaps at three locations below Machu Picchu. We understand Inca bridge construction methods because present-day Quechua Indians still build them from time to time using strong ropes made from long grass. Figure 8-19 shows an old sketch of an Inca bridge.

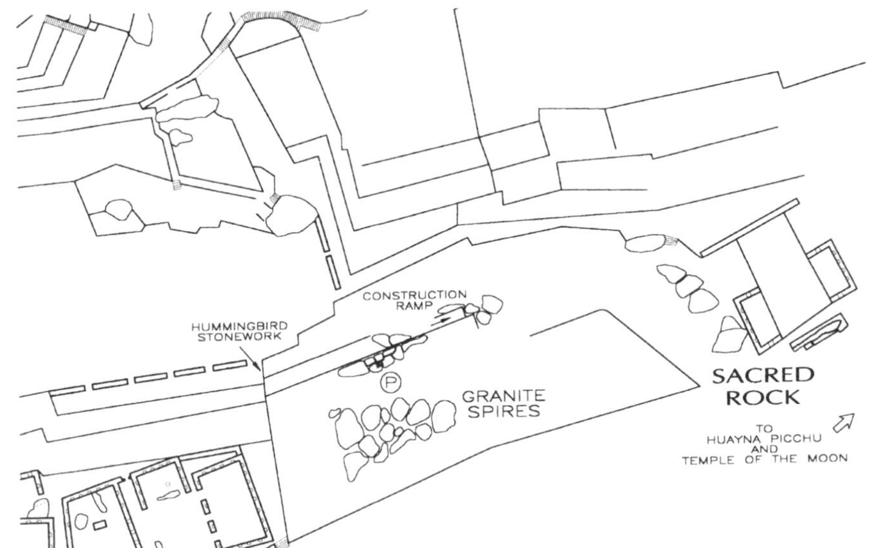

◀FIGURE 8-20
The area of the Granite Spires was never completed. However, it has a petroglyph designated as P, a construction ramp, and an uncompleted wall to the right. To the left, on the outside of the enclosure wall, random stonework forms a hummingbird.

Renegade Inca Stonemasons

The Inca engineers' and planners' wall construction did not include stone art known as zoomorphic iconography. One exception, however, is at Machu Picchu on the megalithic wall south of the Granite Spires as shown in Figure 8-20. Here, eight random stones form the pattern of what appears to be a hummingbird with eggs and a baby hummingbird on its head (Figure 8-21).

Almost unanimously, Inca scholars believe that the stone configuration is purely random. However, an engineer might challenge this opinion on the basis that eight stones randomly forming megalithic art is too much of a coincidence. Coincidence or not, the "megalithic art" is worth a visit. It might be the work of renegade Inca stonemasons who strayed from the strict empire-wide practice.

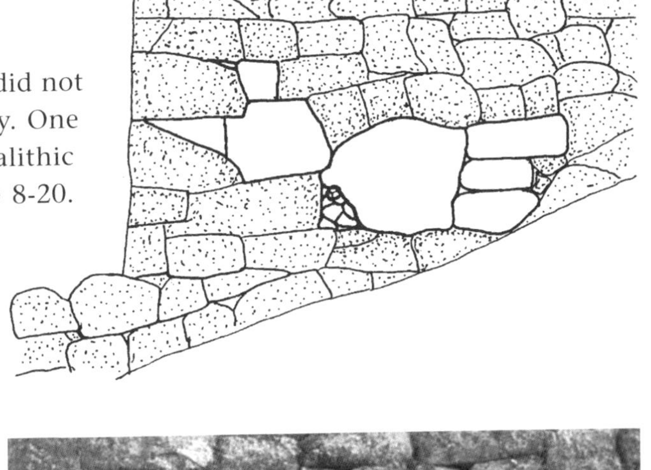

FIGURE 8-21▲
The random placement of stones in the southern enclosure wall of the Unfinished Temple of the Granite Spires has taken on the shape of a hummingbird. If the stones had been purposely placed, the art would be called semeiology.

Tools of the Trade

The excavations at Machu Picchu, carried out by Hiram Bingham and others up to the present time, have identified bronze and stone tools, including levers, chisels, points, axes, hammers, and knives (Figure 8-22).

The bronze tools at this location were the product of combining copper and tin, containing 10 to 13 percent tin. The bronze axes have a trapezoidal blade and the upper part has two horizontal appendices to be attached to a wooden handle. The crowbars are long, relatively thin pieces of a rectangular

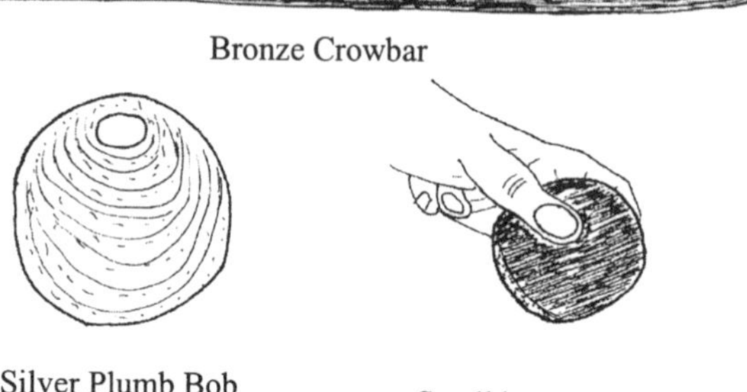

FIGURE 8-22▶
Hiram Bingham discovered construction tools at Machu Picchu in 1912. Clockwise from upper left: bronze knife, bronze axe, bronze crowbar, hammerstone, and plumb bob.

cut, with one end lightly worked and the other rounded. Other bronze tools likely existed.

Hammer stones, abundant in Machu Picchu, have been found in nearly all of the excavations, in dumps, and even in the center of the walls, forgotten by the Inca stonemasons.

Finally, the civil engineer should not pay attention to the modern myth surrounding the quarry stone cutting shown in Figure 8-23. This stone splitting is the work of someone who was emulating Egyptian techniques, not Inca methods.

◀FIGURE 8-23
This stone cutting method is not Incan, but a modern effort emulating Egyptian techniques. Don't be fooled when you visit the quarry.

Chapter Nine

Cultural Background

Inca Heritage

by Gordon McEwan, Ph.D.

In the popular imagination, ancient Peru often consists of images of lost cities, gold, Incas, and Spanish conquistadors. But these images suggest only a small part of the rich history of the Peruvian Andes where many complex societies rose and fell over the course of thousands of years. The peoples that the Spaniards encountered in the 16th century are known from fragmentary written descriptions and chronicles.

Unlike the pre-Columbian Mesoamericans, the ancient peoples of the Peruvian Andes lacked a system of writing, so no pre-conquest native accounts of their history or literature exist. Nevertheless, the products of these civilizations, their magnificent works of art, their architecture, temples and cities, agricultural terraces and irrigation works, and the bones and mummified remains have survived to provide a wealth of information. Through study of these varied remains, archaeologists can reconstruct the prehistory of Andean South America.

Andean Culture

As it was in ancient times, the population of the Andes is comprised of many different ethnic groups. These peoples originally spoke their own languages and had their own distinct cultural practices. The Spanish conquest had a homogenizing effect in that much of the linguistic diversity disappeared. Most of the population of the Andes now speaks one or more of three languages: Quechua, Aymara, or Spanish. While even today the Andean population has much ethnic diversity, certain commonalities bind these people into an Andean culture. Much of this culture is shaped by the universal basic subsistence activities of farming, fishing, and herding. The difficulty of farming the steep Andean slopes and desert river valleys made cooperative efforts essential. With no draft animals, plowing was done by hand. A group of farmers each digging one furrow could plow a field much more efficiently than a single farmer (Figure 9-1). In areas requiring irrigation, construction and maintenance of canals was possible only with large groups cooperating and coordinating their labor. The reciprocal obligation to help neighbors or kinsmen with their labors in return for their help became a key feature of Andean society.

Communal holding of land and its surplus was also an important feature of many Andean societies. In Inca times, the state stored surplus food in warehouses for insurance against times of want (Figure 9-2). In times of famine, this food was redistributed to those in need. On a community level, widows, orphans, the sick, and elderly were taken care of by the surpluses of their kin groups or had their lands worked

◄**FIGURE 9-1**
Huamán Poma, born of Inca nobility, drew this group of Inca plowing a field under a bright sun prior to A.D. 1615. Note the lady bringing drinks.

for them. The storage and redistribution of surpluses, together with farming strategies involving multiple eco-niches, were the main means by which the Andean population could cope with the unpredictability of their environment.

The Andean societies' religious beliefs reflected a deep concern with ensuring fertility and water, and warding off disease. The Inca maintained a hierarchically organized priesthood to serve the temples and shrines of the gods. A separate category of supernaturals was animistic spirits called *huacas* that inhabited everything in nature. They could be found in mountain peaks, unusual natural phenomena, odd-shaped stone outcrops, stone idols, and mummies. *Huacas* were honored with shrines and offerings and taken care of, since an angered *huaca* could cause drought, famine, or other disaster.

Although our knowledge of pre-Inca religious beliefs is limited, archaeological and iconographical studies suggest that the basic concerns with fertility and the agricultural cycle are ancient. Certain animal images—jaguars and pumas, caymans, and raptorial birds—are believed to have been associated with water, fertility, and the supernatural. These images continuously reoccur in the religious art and temples of the ancient Andes.

FIGURE 9-2▲
Mountainside storehouses, or qolqas, still stand throughout the Andes as sentinels to the Inca planning and social organization genius. Foodstuff was stored for poor harvests and to provide nutrients for military legions passing by. The qolqas were symbols of the power of the state.

Treatment of the dead in Andean societies varied through time and between cultures. Archaeological evidence indicates that most ancient Andean peoples believed in an afterlife. Bodies were typically buried with grave offerings of ceramic vessels—sometimes containing food, clothing, ornaments, and other items used in life. High ranking, wealthy individuals were buried with more and better quality objects. It is not uncommon to find mummies wrapped in many layers of well-preserved fine textiles. The arid desert coast of Peru preserved the contents of many of these tombs in near-perfect condition.

FIGURE 9-3▲
The mummy of an ancestor enjoyed the same privileged treatment as during life.

In Andean society at the time of the Spanish conquest, the dead were considered to have considerable influence over events affecting the living. The mummies of the dead Inca emperors were never buried but continued to preside over their households and property as if still alive (Figure 9-3). Dressed, fed, and entertained, these corpses continued to function in society as if they had suffered only a slight change in status. The Inca considered them to be powerful *huacas* and sought their advice on matters of importance. How far back into the past the Inca observed this practice is unknown, but the rich furnishings of many of the well-preserved coastal tombs indicates that the dead were highly regarded and well cared-for even in ancient times.

Andean Technology

Complex societies evolved in the Andes without a well-developed writing system. Instead, records and accounts, history and literature were recorded and remembered through a unique mnemonic device, a type of memory aid called a *quipu* (Figure 9-4). A *quipu* allowed its maker to remember and record a large body of complex information by joining together knotted strings and cords of various colors in a particular order. Typically, Inca *quipus* held statistical records of production, storage, redistribution, as well as census data. Historical information, such as the succession of rulers or famous events, could also be stored on a *quipu*. Quantities were indicated by use of varying types of knots. Additional meaning was relayed through the position of the knot on the cord and through the color of the cord. The device was limited in that only its maker could interpret it. To interpret a *quipu* one has to know in general to what it refers, as well as the specific meanings of the different colors of cord and types of knots.

Other technological innovations in the Andes were governed by the raw materials and resources found in the natural environment. Hard metals such as iron and steel were unavailable, but the softer precious metals, gold and silver, were relatively abundant, as was copper.

FIGURE 9-4▲
An official has two quipus of knotted strings. Quipus were used to keep records of people, storehouse contents, and taxpayers. The Inca had no written language.

FIGURE 9-5▶
This ingenious foot plow was a type of digging stick used to plow fields.

FIGURE 9-6▲
The Inca Trail was built over difficult terrain, yet many sections are remarkably straight.

Bronze, produced by mixing copper and tin, became the hardest metal known to ancient Andean peoples. Good quality clay was also readily available in the Andes, and manufacture of fine ceramics was widespread.

The lack of large, domesticated animals for transportation or work, coupled with the impossibly steep terrain of the highlands, discouraged the development and use of the wheel. With all traffic on foot, roads and bridges assumed a distinctive form. Long, steep staircases are a common feature in pre-historic Andean roads. Narrow, hanging suspension bridges made of rope were well adapted to foot traffic (Figure 8-19). Without draft animals, farm labor was performed by hand. An ingenious foot plow, a type of digging stick, allowed a person or group of people, lined up in a row and each producing one furrow, to plow a field effectively (Figure 9-5). On the coast and shores of the highland lakes where timber for boats was not available, the people learned to build fishing boats of reeds.

Using tools of stone and bronze (Figure 8-22), the ancient Andean became master stone masons, producing great works of architecture involving complex engineering skills without the precision instruments known in the Old World. They built enormous temples, palaces, and fortresses of adobe bricks or stone and laid remarkably straight roads over the most difficult terrain (Figure 9-6). They even sculpted the landscape to suit human needs, constructing agricultural terraces to conform to the topography for aesthetic as well as functional purposes (Figure 5-1).

Inca History

According to Inca legend, a small band of highlanders migrated into the valley of Cusco in the southern Peruvian sierra around A.D. 1200. During the next few centuries, the huge empire of the Inca was to spring from this small group. The Inca claimed that their place of origin was the town of Paccaritambo, a few kilometers (miles) to the southwest of Cusco, where their ancestors had come forth into the world from three caves. This original group was led by the first Inca ruler, Manco Capac, and was comprised of his three brothers and four sisters. After many adventures, Manco led his small band into the valley of Cusco, where they then established themselves by force of arms and brought order and civilization. Other stories held the place of origin to be an island in Lake Titicaca to the south of Cusco, from which Manco led the Incas north to the valley of Cusco. Yet other accounts combined these two legends into one, with the Incas migrating underground from Lake Titicaca to Paccaritambo where they emerged from the caves of origin.

After their arrival in Cusco, the Incas slowly increased their influence through intermarriage and military raids against their neighbors (Figure 9-7). The city of Cusco grew from a late intermediate period settlement, but through the reign of the eighth Inca, it was little more than an ordinary Andean highland town. The turning point in the history of the city and the Incas themselves, was the great Chanca crisis near the end of the reign of Inca Viracocha (A.D. 1438). By this time, the Incas had increased their domain to include the entire valley of Cusco, including the Oropesa and Lucre basins and a large part of the neighboring Yucay valley. A powerful warlike confederation known as the Chanca began to expand to the south, probably from the Ayacucho basin, the earlier Wari imperial seat. They threatened Cusco and

FIGURE 9-7▲
Inca warriors made it possible for the Inca Empire to expand and increase its influence. Armies could be raised because of the ability of the Inca to produce food surpluses.

very nearly defeated the Incas. The Inca Viracocha abandoned the city and fled to the neighboring valley, but at the last moment, one of the royal sons, Inca Yupanqui, rallied the Inca armies and, in a heroic effort, defeated the Chanca forces. After this victory he deposed his father, Inca Viracocha, whose failure to defend Cusco was viewed as a disgrace. Inca Yupanqui took the name Pachacuti and assumed the throne to became the first of the great Inca emperors.

FIGURE 9-8▼
Cusco, the Inca capital, was constructed in the form of a puma (mountain lion) with its head the fortress-temple of Sacsahuaman. Felines, especially the puma, were sacred.

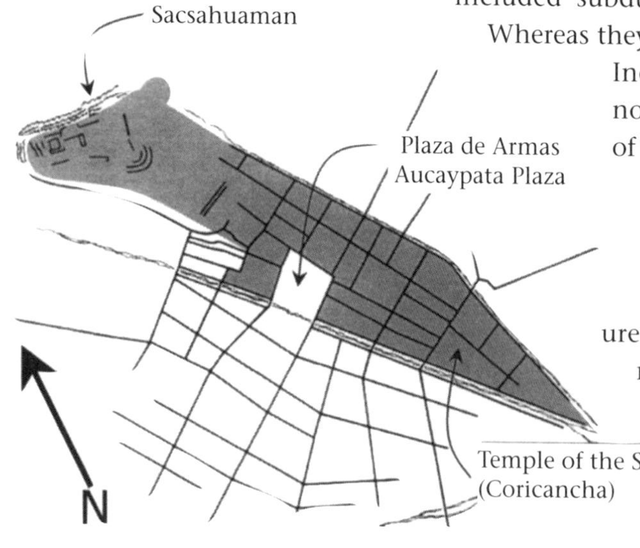

The name Pachacuti, or Pachacutec as it is sometimes given in the chronicles, means "he who shakes the earth" or "cataclysm" in Quechua, the language of the Incas. It was an appropriate name for a man who literally reorganized the Inca world. His first acts as emperor included subduing the neighboring peoples in the Cusco region. Whereas they had previously been associated rather loosely with the Incas, mostly by persuasion and family ties, they were now firmly brought under control as vassals of the lords of Cusco. He then launched a series of conquests that rapidly evolved what had been the tiny Inca domain into an expanding empire. Pachacuti conquered large areas of the sierra, moving north into the central Peruvian highlands and south to the shores of Lake Titicaca (Figure 1-4). He also turned his attention to reorganizing and rebuilding the city of Cusco and designing the empire.

Pachacuti set for himself the task of reconstructing Cusco as a suitable capital for the empire he envisioned. The city was constructed in the form of a puma, incorporating the fortress-temple of Sacsahuaman as its head (Figure 9-8). Like so many New World peoples, the Incas held felines, especially the puma or mountain lion, to be sacred. The body was comprised of residential buildings and palaces laid out in a grid between the Tullumayo and Saphi Rivers. The basic building unit of the city plan was an architectural form called the *cancha* that was comprised of a series of small houses arranged within a rectangular enclosure (Figure 9-9).

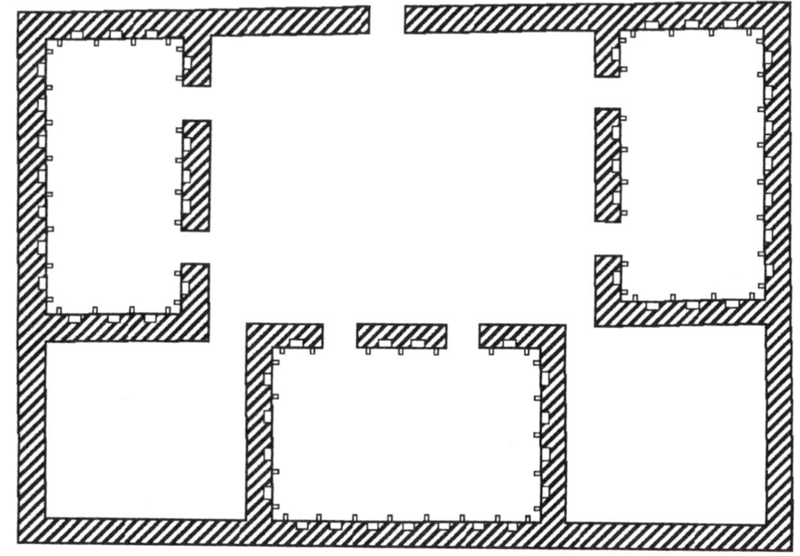

◀FIGURE 9-9
This typical Inca building enclosure is often referred to as a cancha. *Note the geometric perfection of doors and wall niches.*

The city of Cusco was conceived as the center of the empire (Figure 9-10) where its four quarters came symbolically and physically together. Four highways coming from each of the four quarters, or *suyus*, converged in the great central plaza of the city. From this four-part division the empire took its name of Tawantinsuyu, meaning "the land of four quarters."

In addition to rebuilding Cusco (Figure 9-11), Pachacuti also initiated building projects in the environs of Cusco and on his royal estates in the Urubamba Valley. The most famous of these is the so-called "lost city" of Machu Picchu, but he also built royal estates at Ollantaytambo, Patallacta, and many smaller sites in the valley.

Pachacuti initiated other building projects such as the famous royal highway of the Inca (Figure 9-12), which facilitated communication within the expanding empire and provided a means of rapidly moving the army to wherever it was needed. Following and expanding the routes of the old highways of the earlier Wari Empire, standardized highways, often walled and paved, linked the various regions of the growing empire to Cusco. Storehouses (*qolqas*) and rest stops (*tambos*) were built to serve the army as it marched. A system of relay runners (*chasqui*) formed an effective postal system to transmit verbal messages and instructions. Pachacuti and his successors also built towns and provincial administrative centers in the various conquered territories as the empire expanded.

FIGURE 9-10▼
Original Inca stonework in the Coricancha at Cusco was perfectly fitted. Cusco was the center of the Inca Empire.

FIGURE 9-11▶ *Perfectly fitted Inca stones of this Cusco building wall have withstood many earthquakes.*

Pachacuti's son, Topa Inca, succeeded him as emperor in A.D. 1471 and continued to expand the empire. Topa Inca moved the imperial frontier north into what is now Ecuador and south into what are now Bolivia, northern Chile, and northwestern Argentina. By A.D. 1476 he had achieved the conquest of the Chimu Empire on the Peruvian north coast—the last serious rival for total control of the Andean area. Topa Inca reigned until A.D. 1493 and was succeeded in turn by his son Huayna Capac.

Huayna Capac continued to expand the boundaries of the empire to the north and east, incorporating much of what are modern Ecuador and the northeastern Peruvian Andes. Compared to his father, however, his conquests were modest. Huayna Capac spent so much time on his difficult northern campaign that the social fabric of the empire became strained. He was absent for many years at a time. Surrogates had to stand in for him at important festivals and ceremonies, and the people of Cusco began to feel out of touch with their emperor. A new and potentially rival court even grew up around him at his northern headquarters at Tomebamba in Ecuador. Administratively the empire had become difficult to govern. Decisions from the emperor took a long time to reach Cusco and even longer to be disseminated to the rest of the empire. Controlling the far-flung outposts of the empire became increasingly difficult.

A severe crisis finally came when Huayna Capac suddenly died of what may have been smallpox in A.D. 1527. The disease, introduced by Europeans arriving in the New World, preceded the Spanish conquistadors as they journeyed across South America. Thousands died in a very short space of time, including Huayna Capac's appointed heir, who survived his father by only a few days. The confusion about the succession created even more strain on Inca society, and finally a civil war broke out between two brothers who were rival claimants for the throne. One of the brothers, Huascar, had succeeded to the throne in Cusco in A.D. 1527. Atahuallpa, who had been with his father and the imperial army in Ecuador at the time of Huayna Capac's death, chal-

lenged him. A large part of the army rallied behind Atahuallpa, and a bloody war ensued. Taking the city of Cusco in A.D. 1532, the forces of Atahuallpa eventually prevailed. The emperor Huascar was captured and imprisoned.

As Atahuallpa moved south to Cusco with a large army, he was met at the town of Cajamarca in the northern highlands by the Spanish forces led by Francisco Pizarro. In a stunning surprise move, Pizarro and his small band of 168 men attacked and captured Atahuallpa in the midst of his huge army. Pizarro held the emperor captive for nearly eight months, waiting for the ransom that would secure Atahuallpa's release. The emperor offered to fill a room once with golden objects and twice with silver. This treasure chamber measured 6.7 meters (22 feet) by 5.2 meters (17 feet) and was filled to a height of over 2.4 meters (8 feet). In all, almost 10 metric tons (11 tons) of treasure was collected throughout the empire and sent to Cajamarca. While he was in captivity, Atahuallpa secretly sent orders to have the Inca Huascar killed. He eliminated his rival, but he himself was killed by the Spaniards shortly thereafter in July 1533. With the death of Atahuallpa, the last of the independent Inca rulers had fallen. The Incas continued to resist the Spanish for many years, but the Inca Empire ceased to exist.

FIGURE 9-12▲
This segment of the Inca Trail at Machu Picchu is part of the famous royal highway of the Inca. Hydrogeologist Gary Witt is descending a granite stairway, which is a common trail feature.

Inca Society

Most of what is known of Inca society is based on the Spanish chronicles, some of which were eyewitness accounts. Inca history viewed the great emperor Pachacuti as the founding genius of the Inca State. Pachacuti's reconstruction of the Inca capital coincided with a complete reorganization of Inca society. At the apex was the emperor himself, called the Sapa Inca, and the noble families of pure Inca blood. There were never more than about 500 adult males, and perhaps 1,800

◄FIGURE 9-13
This military man held not only his shield and mace, but also a cup for drinking the Inca beer made from corn.

people in all who carried pure Inca blood. This lineage or extended family owned the empire. All the important governmental posts, the governorships of each of the four quarters of the empire, the army, and the religious institutions were held by pure-blooded Inca. Below them were the Incas by adoption, or Hahua Incas, comprised of neighboring peoples held in high enough esteem by the pure blood Inca to be trusted with important positions when there were not enough royal Inca to fill these posts. Below these were provincial nobility, who were local ethnic lords confirmed in place by the Inca administration. At the bottom of the social pyramid were the Hatun Runa (big men), who were the common heads of households. Hatun Runa were decimally organized in groups of 10, 50, 100, 1,000, 5,000, and 100,000 families for administrative purposes. Each decimal division had an official responsible for its administration. This organization was the key to the success of the empire. The family was the basic taxpaying unit. Each family provided a set amount of labor or service to the state rather than wealth in the form of material goods. The state would, in turn, use this labor to generate wealth through the production of goods, cultivation of lands, construction projects, or military conquest of new territory.

The social organization of the empire was based on a complex series of reciprocal obligations between the rulers and the ruled. The Hatun Runa paid taxes to the imperial government in labor service. In return, the government provided social services to protect the population in times of want and natural disaster. Food and other goods were collected and stored for use during drought or famine. The government set aside some income from government lands for widows and orphans. The government also provided maize beer and food for ritual feasting on annual holidays (Figure 9-13). The imperial government ensured that every citizen was fed and clothed.

The labor tax was collected by drafting work parties to build state projects such as roads, buildings, and agricultural terraces. This labor also worked government-owned lands to produce surplus food for storage and to support the nobility as well as priests and women of the religious institutions. Sometimes citizens paid the labor tax by serving in the armies or by producing goods such as wool or cloth for the state.

Pachacuti also organized Inca religion into an imperial institution. The major gods of the various peoples incorporated into the empire were all included in the Inca pantheon and appropriate temples and shrines for them were built and maintained. In addition to Inti the sun god, the Inca patron, the major gods were Illapa the god of thunder, Pachamama the earth mother goddess, Mamacocha the sea goddess, and Mama Quilla the moon goddess. Above all was Viracocha, the great creator deity of the Andean peoples. In a separate category were the *huacas*, whose specific manifestations occurred in mountain peaks, unusual natural phenomena, odd-shaped stone outcrops, mummies, and stone idols. Huacas were physically arranged along imaginary lines called *ceques* radiating outward from the Coricancha, the temple of the sun, in Cusco (Figure 9-14).

FIGURE 9-14▲
The curved wall of the Coricancha, the Temple of the Sun, in Cusco, has withstood many earthquakes. It was the center of the Inca Empire.

Inca religion emphasized ritual and organization rather than mysticism or spirituality. The priests, as well as the *huacas*, were arranged in a rigid hierarchy. Religious rites chiefly focused on ensuring the food supply and curing disease. Divination was also important. The Inca maintained an elaborate ritual calendar of public ceremonies and festivals, most associated with stages in the agricultural cycle such as plowing, planting, and harvesting. Others were related to solstice observations, puberty rites, and new year celebrations. Some festivals and ceremonies were held only on special occasions, such as coronations and military victories, or to counter drought and disasters. Almost all rituals involved elaborate public ceremonies

FIGURE 9-15▲
The Inca made offerings, sometimes of coca leaves, shells, or figurines, at huacas.

that included dancing, reciting, drinking alcoholic beverages, and making sacrifices. Usually llamas or guinea pigs were sacrificed. On very special occasions war captives or young boys or girls were sacrificed—strangled or killed with special elaborate knives called *tumi*. Offerings of figurines, such as small stone alpacas or llamas and gold and silver figurines, as well as shells and coca leaves were commonly buried in the ground around *huacas* (Figure 9-15).

To institutionalize the cult of the royal mummy, Pachacuti invented or perhaps reorganized an institution called the *panaca*. A *panaca* was comprised of the royal household and descendants and was headed by a second son of the emperor. Each emperor formed his own *panaca*, which continued to function after he died and continued to hold all of his wealth. The *panaca* cared for the emperor's mummy and managed his estates and wealth in perpetuity. His successor thus inherited the empire but not the accumulated wealth of his predecessors. Unfortunately, the *panacas* de-stabilized the empire since the dead emperors continued to wield influence through their *panacas*, and eventually a great part of the wealth of the empire was controlled by the *panacas* rather than by the head of state.

Inca society was highly stratified, and upward mobility was limited. The only way a person could improve his or her position was by being a successful warrior, by being attached as a type of servant to an important noble household, or by being chosen as an *aclla* (chosen woman). The state controlled most aspects of its citizens' lives, and a strict code of law was applied even more harshly to the nobility than the commoners. Travel and dress were also strictly regulated; no one could move about the empire or change from his native costume without the state's permission. The basic social unit beyond the immediate biological family was a kin group called the *ayllu*. Members of the *ayllu* held land communally and often made decisions collectively. Rights to land and water were inherited through membership in the *ayllu*.

Because the rights of the *ayllu* were based on ancestral claims, worship of specific ancestors was an essential means of legitimizing land

tenure. Ancestors were often kept as mummies or identified as specific *huacas*. They were clothed, "fed," and provided with drink, as were royal mummies who performed this function for the royal Inca *ayllus*. The ancestors were consulted on all important matters. Ancestors also had vested in them the responsibility for ensuring that sufficient water was available to irrigate the land. The water was provided in the form of rain, rivers, and springs. One method the Inca used to control their subjects was to physically control their ancestors, who in turn gave them control over land and water rights. Ancestors and *huacas* were brought to Cusco as hostages and held to ensure the good behavior of conquered peoples.

FIGURE 9-16▲
By physically altering the landscape, the emperor could express his power and divine will as shown here at Pisac, where the steep hillsides are gracefully shaped.

Many of the above-mentioned aspects of Inca culture are expressed in the royal estate of Machu Picchu. This site was built by the greatest of the Inca emperors, Pachacuti, and was owned by him and his *panaca* until the collapse of the empire. We can imagine that even after his death he continued to visit the site and to oversee its management.

Machu Picchu served many functions that are comprehensible only in terms of Inca culture and belief systems. As a divine ruler, the emperor could express his power and divine will by physically altering the sacred landscape (Figure 9-16). He built his estate not on the mountain, but within it and its sacred environment. This was meant not only to impress his mortal subjects, but also to demonstrate his power among the other gods and to tie himself closely to their supernatural power. The vast majority of his subjects never saw Machu Picchu, nor were they probably aware of its existence.

Several key aspects of Machu Picchu's location stand out. It was not built at a crossroads or gathering place but is remote and surrounded on three sides by the waters of the Urubamba River (Figure 9-17). This river is a major *huaca* embracing the site. The site also approaches the sky within the sight lines of several major mountain peaks, another category of powerful *huacas*, which renders it an appro-

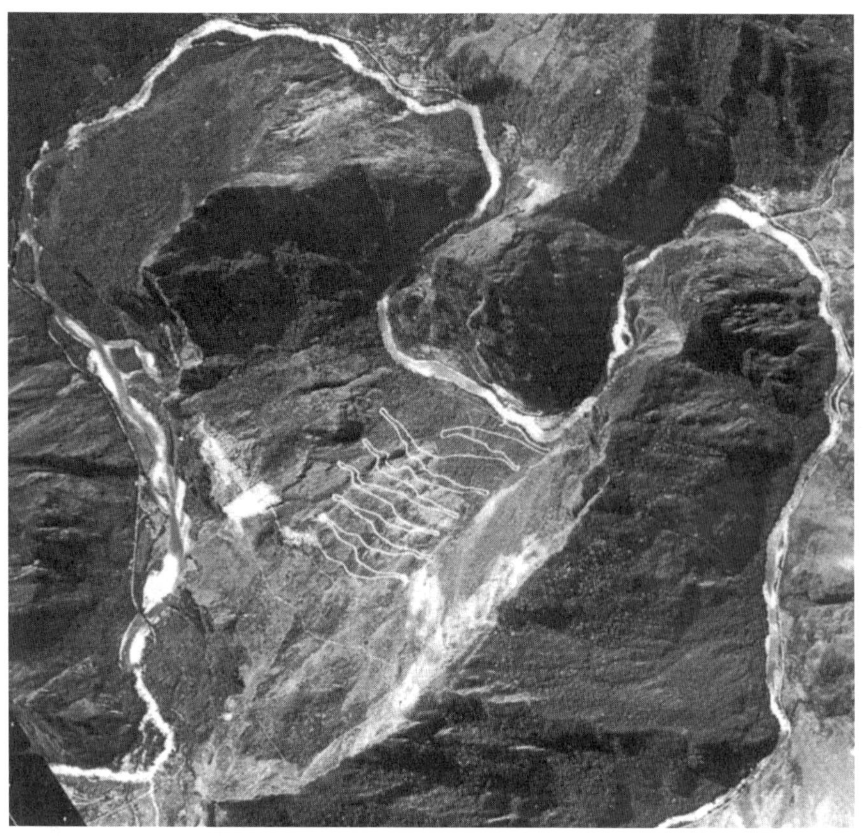

priate place to worship mountain spirits. Within the site itself, are numerous sacred stone outcrops. These *huacas* are generally tied into the architecture of the site and some are actually enclosed in buildings specially constructed for this purpose (Figure 9-18). Many are also aligned with the sacred mountain peaks and in some cases are carved to echo the contours of the distant mountain on which they are focused. A canal system brought water to the site, carrying it through a series of elaborate fountains and then allowing it to drain into the river below. On one level this system provides a practical water supply for the inhabitants of the site, but it also recapitulates the sacred water cycle regulated by the ancestors (Figure 9-19). Water falls from the sky into the cachement

FIGURE 9-17▲
With the waters of the Urubamba River on three sides, Machu Picchu was embraced by a major river, a huaca. *Major mountain peaks in the line of sight meant that this was an appropriate place to worship both water and mountain spirits. The switchbacks make up the Hiram Bingham Highway leading to Machu Picchu. The river flows from right to left.*

FIGURE 9-18►
This rock huaca *is enclosed by the Temple of the Sun at Machu Picchu. It was enclosed within a building specially constructed for that purpose.*

FIGURE 9-19▲
The fountains of the Inca recapitulated the sacred water cycle regulated by the ancestors. This fountain is at the Tipon ruin, downstream from Cusco.

basin, is collected into the canal system from which it nourishes human life, then flows into the river and ultimately the sea. There it evaporates back into the atmosphere to begin the cycle again. Ultimately, Machu Picchu was a place where the divine emperor could commune with his fellow deities and participate in the great cycle of life, land, and water. Balance and harmony between the real world and supernatural world were expressed and maintained.

The Inca Empire, although short lived, was the culmination of thousands of years of Andean civilization. From their predecessors they inherited a body of statecraft and much of the physical infrastructure for the empire. They also inherited and participated in a great Andean culture based on religious ideas that were thousands of years old. That they were the inheritors of this great tradition does not in any way diminish their own achievement, however. It was the peculiar Inca genius for organization that allowed them to make profitable use of their cultural inheritance. Of the late Andean societies, they alone were able to weave together the disparate elements of the many Andean cultures through military prowess, ideology, and extraordinary statecraft by drawing on thousands of years of cultural inheritance. In terms of geographical extension, military power, and political organization, the Inca created the greatest of the pre-Columbian empires (Figure 9-20).

FIGURE 9-20▲
The Inca created the greatest of the pre-Columbian empires using military might and political organization. They were successful because of their ability to produce excess food, enabling them to free-up sizable portions of the population for military service and public works.

To walk through Machu Picchu is to walk in the footsteps of remarkable people who revered the mountains and rivers surrounding this mountaintop royal estate. Experience all of Machu Picchu, not just the civil engineering works of these ingenious ancient Native Americans. Don't rush. Absorb what you see and note the details, for the miracle of Machu Picchu is not only in its grand vision, but in its meticulous order and precision. While you can complete the basic tour from the Inca Canal to Conjunto 17 in about four hours, retrace your steps to see additional details. If you stay overnight you can take one or more of the side trips described in this chapter.

Chapter Ten

A Walking Tour

Machu Picchu for Civil Engineers

by Ruth M. Wright

The Main Tour

Seeing the Inca Canal

Whether you are an armchair traveler or actually on site, use the map on pages 96 to 99 in conjunction with this tour description.

Proceed through the ticket entrance gate and head for the storehouses (Figure 10-1). Turn right, down the trail, to a set of stairs. The path then turns left to go between the storehouses. Stay on the main tourist route, traveling north into the heart of Machu Picchu. After passing through the agricultural sector, walking past llama pens on your left, and seeing the curved wall of the Temple of the Sun (Figure 10-2) directly ahead, you will come to the main drain or dry moat (Figure 5-5). Turn left and climb the stairway past nine terraces until you come to the ancient Inca water supply canal (Figure 10-3).

Here you can see the canal as it passes through the urban wall. In 1912, when Bingham's civil engineers and surveyors mapped Machu Picchu, the canal crossed the main drain as a stone aqueduct (Figure 4-9). Notice the small cross-section of the canal and its carefully placed stone lining and controlled grade. Immediately above and to the west is the interceptor surface drain with an orifice that discharges to the main drain (Figure 5-18).

Now walk upstream (south) along the canal. Approximately 42 meters (140 feet) upstream, the canal is on its own narrow terrace wall on its own ledge for about 31 meters (102 feet) where the slope is only about one percent (elsewhere the canal slope averages three percent).

◄FIGURE 10-1
From the ticket entrance gate you see five buildings that are associated with the agricultural sector. This two-story grain warehouse is the uppermost of the five buildings.

FIGURE 10-2▼
The curved wall of the Temple of the Sun is centrally located in the west urban sector. The curved wall encircles an important huaca *carved from the living rock.*

Continuing upstream along the canal, you can see the five storehouses directly ahead. Take a moment to explore the uppermost two-story storehouse and note how it was integrated into the terraces (Figure 4-6). Just beyond the storehouses, the canal makes a near-right-angle bend to the left where a trail crosses the canal. From here the area is closed to the public, but the canal extends upstream another 550 meters (1,804 feet) to the spring on the north slope of Machu Picchu Mountain. The canal right-of-way passes through the outer wall and hugs the steep mountainside on a narrow terrace. Take the trail upward to your next objective, the Guardhouse (Figure 10-4).

Visiting the Guardhouse and Ceremonial Rock

After walking 10 minutes up the trail and through the edge of the rain forest via switchbacks, you arrive at the Inca Trail and a large flat rock platform that provides your first great view of Machu Picchu. From here you can go up the Inca Trail or down to

FIGURE 10-3►
The Inca water supply for Machu Picchu was carried in this canal and through the urban wall shown in the left background.

MAPS

Civil Engineer's Map of Machu Picchu. Courtesy of the Peruvian National Institute of Culture and prepared by the Machu Picchu Paleo-hydrological Survey, Wright Water Engineers.

Sources: Bingham 1930; Peruvian National Institute of Culture, 1987, "Proyecto Senalizacion: Conjunto Arqueologico de Machu Picchu"; Peruvian Tourism Corporation, 1966, "Plano de la Ciudadela Inca de Machu Picchu"; Valencia Zegarra and Gibaja Oviedo 1992; Wright Water Engineers, Inc., 1994 to 2000, proprietary field surveys and observations.

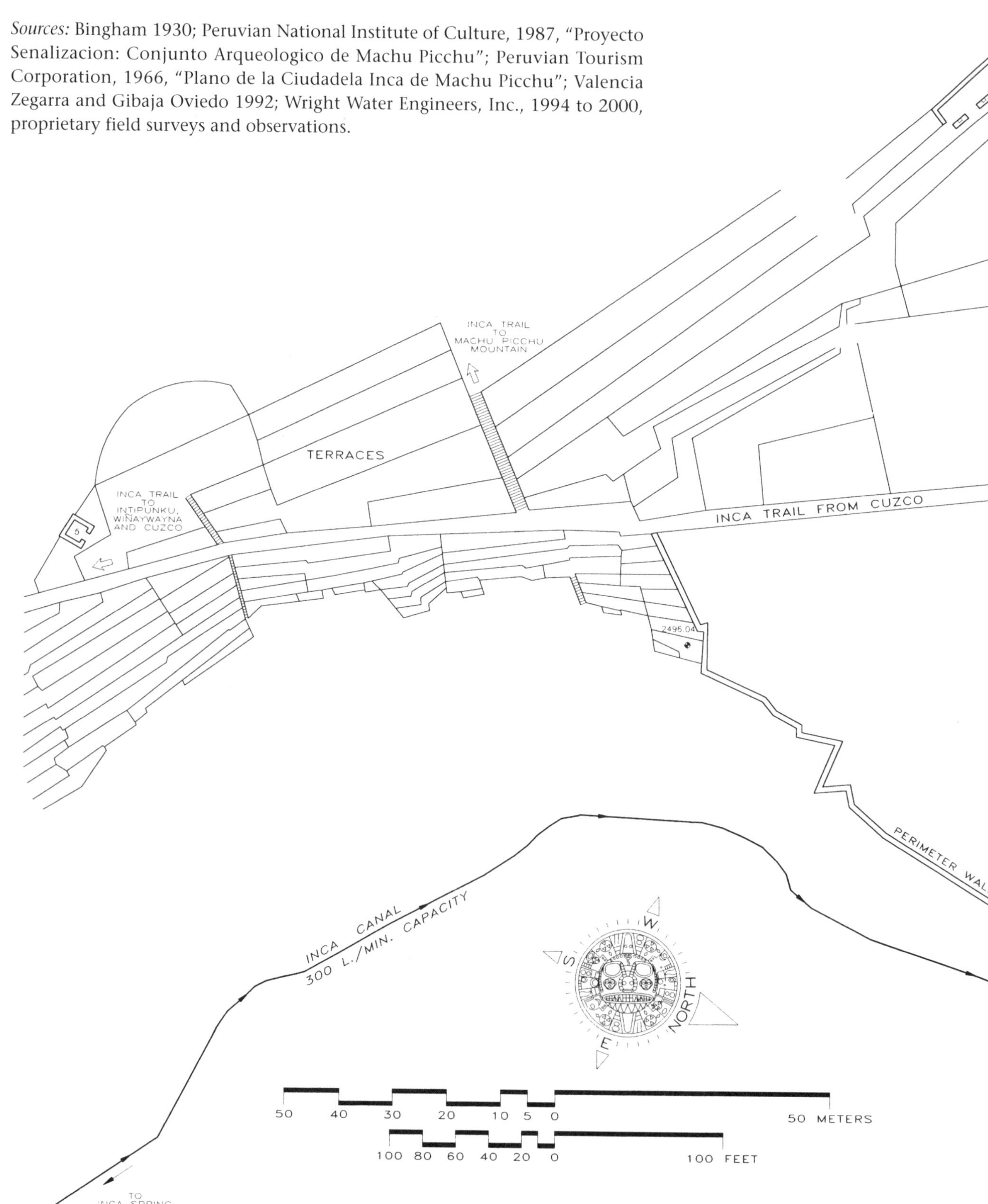

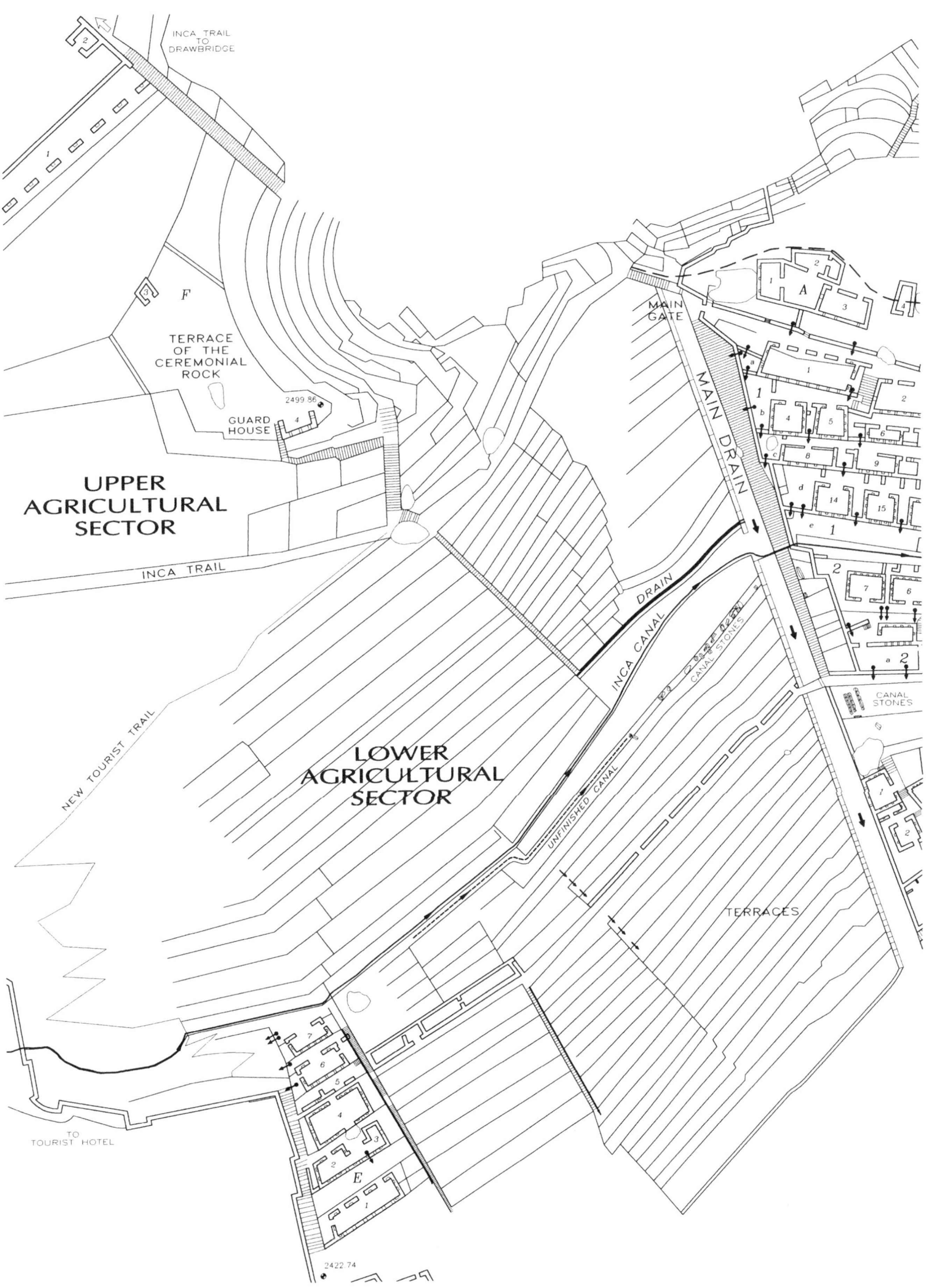

INCA TRAIL TO DRAWBRIDGE
F
TERRACE OF THE CEREMONIAL ROCK
2499.86
GUARD HOUSE
UPPER AGRICULTURAL SECTOR
INCA TRAIL
MAIN GATE
MAIN DRAIN
A
DRAIN
INCA CANAL
CANAL STONES
CANAL STONES
NEW TOURIST TRAIL
LOWER AGRICULTURAL SECTOR
UNFINISHED CANAL
TERRACES
TO TOURIST HOTEL
E
2422.74

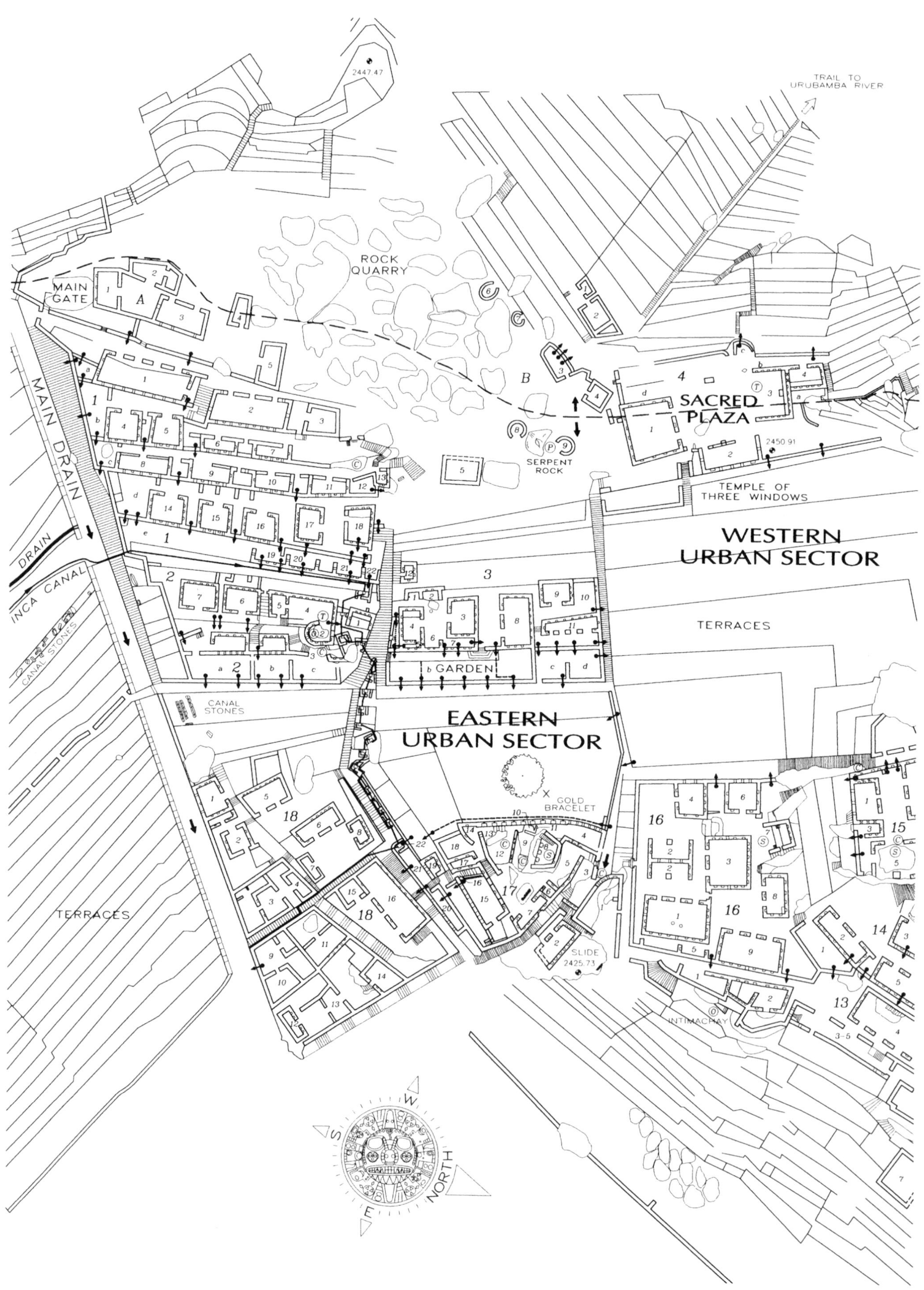
TRAIL TO URUBAMBA RIVER
ROCK QUARRY
MAIN GATE
SACRED PLAZA
SERPENT ROCK
TEMPLE OF THREE WINDOWS
WESTERN URBAN SECTOR
MAIN DRAIN
DRAIN
INCA CANAL
CANAL STONES
TERRACES
GARDEN
CANAL STONES
EASTERN URBAN SECTOR
GOLD BRACELET
SLIDE
INTIMACHAY
TERRACES
NORTH

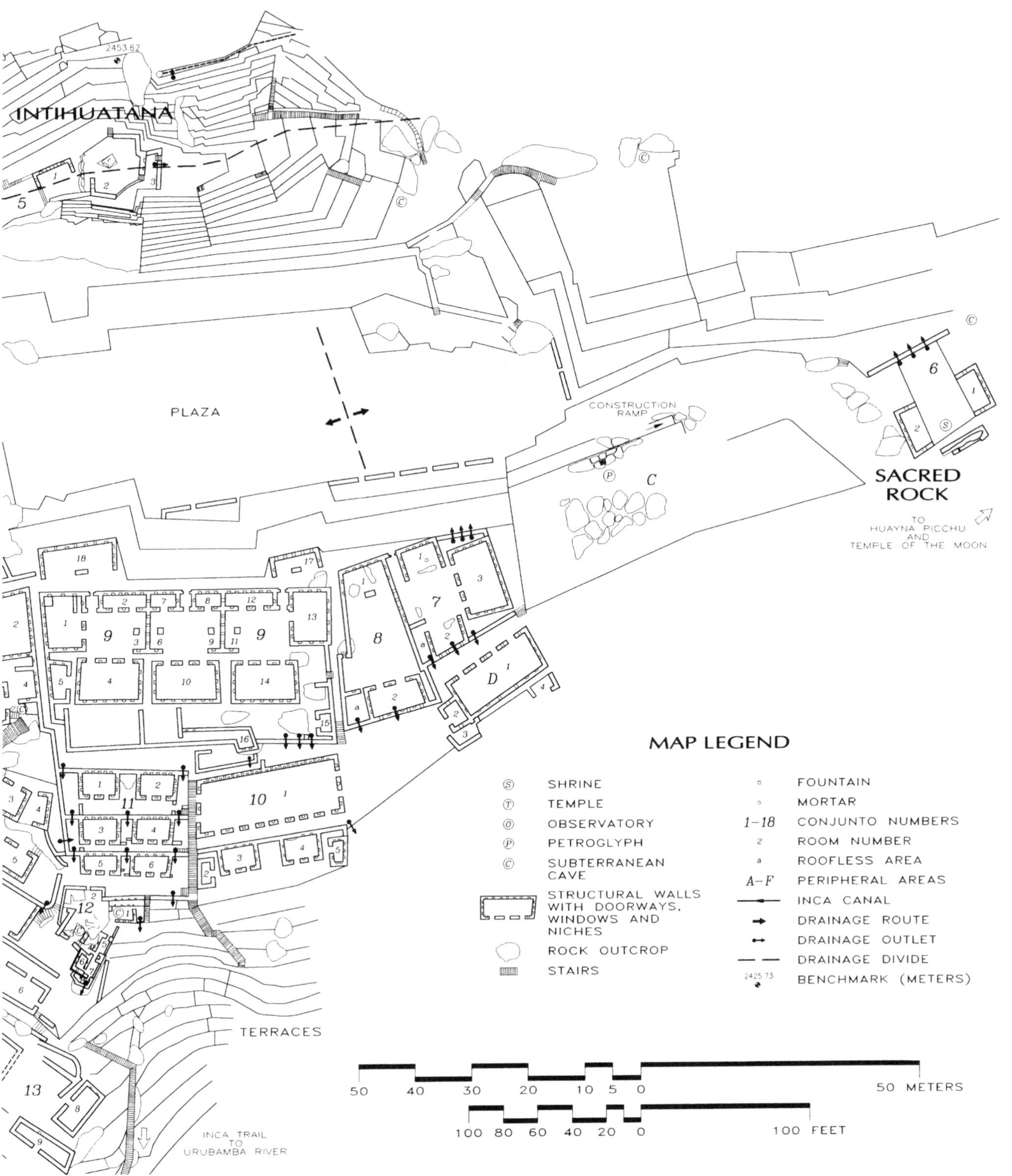

INTIHUATANA
2453.62
PLAZA
CONSTRUCTION RAMP
SACRED ROCK
TO HUAYNA PICCHU AND TEMPLE OF THE MOON
TERRACES
INCA TRAIL TO URUBAMBA RIVER
MAP LEGEND
SHRINE
TEMPLE
OBSERVATORY
PETROGLYPH
SUBTERRANEAN CAVE
STRUCTURAL WALLS WITH DOORWAYS, WINDOWS AND NICHES
ROCK OUTCROP
STAIRS
FOUNTAIN
MORTAR
1–18 CONJUNTO NUMBERS
ROOM NUMBER
ROOFLESS AREA
A–F PERIPHERAL AREAS
INCA CANAL
DRAINAGE ROUTE
DRAINAGE OUTLET
DRAINAGE DIVIDE
2425.73 BENCHMARK (METERS)
50 40 30 20 10 5 0 50 METERS
100 80 60 40 20 0 100 FEET

FIGURE 10-4▶
The three-sided Guardhouse stands at the junction of two trails and overlooks Machu Picchu from a high vantage point. Note the "flying stairs" on the right. Author Ruth Wright is shown on the left for size comparison.

FIGURE 10-5▲
The Ceremonial Rock near the Guardhouse represents exquisite stone carving. Nearby are numerous rounded river cobbles that had religious significance.

the Sacred (Main) Gate. Instead, go up the stairs to the Guardhouse and the Terrace of the Ceremonial Rock (Figure 10-5).

Laid out before you is the grand design of Machu Picchu: the engineered agricultural terraces, the urban sector divided into western and eastern sectors by the wide plaza, and the structures utilizing and enhancing this spectacular and challenging site (Figure 10-6). Huayna

FIGURE 10-6▶
From the Guardhouse, Machu Picchu spreads out before your eyes with Uña Picchu in the left background and the holy mountain of Huayna Picchu on the right.

◄FIGURE 10-7
Leading up to the kallanka, *the "flying steps" built into the terrace walls made it easy to quickly move from one level to another.*

◄FIGURE 10-8
This alternate Inca Trail was controlled with a drawbridge-like structure with logs that could be removed easily.

Picchu Mountain is a stunning backdrop, as are the surrounding mountains. To visualize what Machu Picchu looked like from here in Inca times, imagine buildings with thatched roofs.

One of the larger buildings in Machu Picchu lies on a higher terrace beyond the Ceremonial Rock. It was a great house, or *kallanka*, with eight entrances facing the terrace of the Ceremonial Rock. You can also see that different levels of terraces could be accessed by using the so-called "flying steps," steps protruding from the terrace walls (Figure 10-7).

Look up the Inca Trail to Intipunko (Gate of the Sun), a final checkpoint before visitors descended to the city. A few terraces below the Guardhouse, an alternate trail leads to the Inca drawbridge built on the slip-face of the Machu Picchu fault (Figure 10-8).

Dozens of well-constructed agricultural terraces come into view as you head down to the Sacred Gate. Just before you get to the Sacred Gate, you see the main drain, heading straight downhill, which lies over a minor fault. To the right of the main drain was an earth slide, which the Inca engineers were able to correct. A sharp rock juts up along the staircase about one-third of the way down. It likely was used as a survey marker (Figure 10-9).

Descend the stairs that lead to the impressive Sacred Gate with its huge lintel. The Sacred Gate frames a nice view of Huayna Picchu (Figure 10-10). On the inside of the gate, a stone ring protrudes just above the lintel, and barholds are recessed into the doorjambs on both sides

FIGURE 10-9▲
A sharp rock juts up from the north wall of the Main Drain. It likely served as a control point for field surveys.

◄**FIGURE 10-10**
The Sacred (Main) Gate was constructed with large well-fitted stones and oriented to provide an impressive framed view of Huayna Picchu.

FIGURE 10-11►
The Sacred Gate has a massive lintel beam, a stone ring above, and two barholds. This view looks to the outside.

a few feet above the floor (Figure 10-11). The sketch by Bingham's team shows a likely closure utilizing these features (Figure 10-12). Proceed into Conjunto 1.

A *conjunto* is a grouping of buildings and enclosures with limited entrances, often surrounded by a wall. As you can see from the map, Machu Picchu was planned as a series of separate *conjuntos* (numbered 1 through 18) with paths or stairways leading from one to the other. Conjunto 1 was on five different levels and, besides the storehouses, probably housed workshops.

Proceed down the ramp to a sharp right turn and go through your first double-jamb doorway—only the lower part remains (Figure 10-13). A double-jamb doorway announced the entrances to special places. This one appears to signify a special exit, that is, the final ramp leading to the Sacred Gate.

FIGURE 10-12▲
Bingham's sketch shows how the Sacred Gate interior could be closed using a gate hanging from the stone ring above and secured using the two barholds on either side of the doorway.

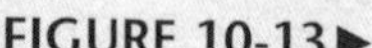

FIGURE 10-13►
Double-jamb doorways announced the entrance to a special place. This doorway apparently signaled a special exit, the final ramp leading to the Sacred Gate.

FIGURE 10-14▶
In the foreground, a rock quarry lies south of the Sacred Plaza. Beyond are the Intiwatana and Uña Picchu. Broken and fractured rock is a result of geologic faulting.

On the left is the extensive granite rock quarry. Ahead is the Sacred Plaza and beyond lies the Intiwatana (Figure 10-14). But, we will first head part way down the Stairway of the Fountains. Go down to, and enter, the lowest level of Conjunto 1 at Rooms 19 to 22, so you can look over the stone wall and down at Conjunto 2.

Going to the Temple of the Sun, Royal Tomb, and Conjunto 2

Straight down from your vantage point on the lowest terrace of Conjunto l, you see the Inca canal on its own narrow terrace (Figure 10-15). To the right, the canal comes through the urban wall and flows to Fountain No. l directly below and to your left.

Directly below is the Temple of the Sun. The Temple sits on a natural rock formation. Notice that the carved rock fills up most of the space inside (Figure 10-16).

FIGURE 10-15▶
The Inca canal runs on this terrace, from back to foreground, but in this 1912 Bingham photograph it is covered with vegetation. Bingham used this photograph as part of his classical composite figure published in 1930.

FIGURE 10-17▲
This single original entrance to the Temple of the Sun is in the same structural condition as when discovered by Hiram Bingham in 1911.

FIGURE 10-16▲
A close-up view of the huaca *at the Temple of the Sun shows the window through which the sun's rays are split in half during the June solstice by the longitudinal stone-cut face in the center of the rock. Thus, the temple is a solar observatory.*

This was a sacred rock, honored and protected by Inca masonry. To the left of the Temple of the Sun is a thatched-roof *wayrona* (a three-sided building), an integral part of this complex. The path between the Temple of the Sun and the canal terrace leads along a long, straight wall that Bingham called "the most beautiful wall in America" because of its masterfully placed, finely cut stones (Figure 8-2).

Go down to the next level through the entrance to Fountain No. 1 immediately on the right (Figure 2-7). All fountains have the same basic design: a small channel to concentrate the jet of water into a shallow rectangular basin, side walls with niches, an entrance for people to fill their *aryballos* (water pots) (Figure 3-7), and an orifice outlet for water to flow to the next fountain. Each fountain has its own spe-

◄FIGURE 10-18
This 1912 Bingham photograph of the entrance to the Temple of the Sun shows the ring above the doorway and the barhold to the right of the Indian boy. (See Figure 10-17 for a comparison to present day.) This photograph was provided by National Geographic Magazine *from their original glass plate archives.*

cial features, this one being a curved canal cut into the rock leading under the path over to Fountain No. 2 between the *wayrona* and the "most beautiful wall." The wall leads to a complete and particularly fine double-jamb doorway. Here the double-jamb is on the outside, the usual location, announcing the special place beyond. Built in the traditional trapezoidal shape of Inca doorways, windows, and niches, it is topped by a lintel stone. On the inside, a stone ring protrudes above the door and barholds are located on each side, like at the Sacred Gate, indicating that it also had a door closure (Figure 10-17). Compare with Hiram Bingham's 1912 photograph (Figure 10-18).

A one-piece rock stairway takes you down. Immediately on the left is the lower level of the Priest's House (No. 5), where you can still see the inside ledge that held up the wooden floor of the second level. Inca buildings rarely had inside stairways, so the outside stairway leading up to the Temple of the Sun would also have been used to access the second floor.

The Temple of the Sun looms ahead with its curved wall, showing the south window with its four stone protuberances, which perhaps

FIGURE 10-19►
The Royal Tomb (mausoleum) lies under the Temple of the Sun. The exquisitely shaped stone and entrance announce its importance.

FIGURE 10-20▼
From inside the Royal Tomb, the carefully shaped stone provides a fitting frame for the view of the valley beyond.

were used for astronomical/observatory purposes (Figure 7-4). The engineers did a remarkable job in connecting the temple to the natural rock formation to create a unified whole. The wall leans inward for stability, and the stone layers decrease in height from bottom to top.

Beneath the temple is the Royal Mausoleum (Figure 10-19). The exquisitely carved stepped white stone at the entrance, the low stone on the floor carved into a shape similar to the Intiwatana stone, the unusually large niches, and the stones carefully fitted between the natural rocks in an hourglass shape exemplify the genius of Inca engineers to meld natural and man-made features into an aesthetic whole. The view from inside accentuates the sculptured stone (Figure 10-20).

Now go out directly to the Stairway of the Fountains and down a few more steps. The Staircase of the Fountains is divided here, making way for Fountain Nos. 5 and 6, an inspired combination (Figure 10-21). From near this spot you can also see the special drainage channel that removed rainwa-

FIGURE 10-21►
The Stairway of the Fountains bifurcates at Fountain Nos. 5 and 6. The cut-stone channels are designed for a maximum of 100 liters per minute. The open side of the wayrona *is at the top.*

ter from the front door of the Royal Residence on the north side of the stairway.

Go up the steps to the *wayrona*, with one side open to a spectacular view of the valley, the mountains, and any ceremonies that were performed here. Modern-day chalk survey lines on the walls inside the *wayrona* and other buildings are used to track settling of the structures. The topographical map shows geological faults, and the entire site is susceptible to earthquakes (Figure 1-6).

A bypass channel (likely modern) runs along the base of the *wayrona* from just above Fountain No. 3 and discharges directly into Fountain No. 4. Today, water that is diverted into Fountain No. 3 does not reach Fountain No. 4. Settlement of the rock formations has opened up routes for water to flow down into cracks and fissures at Fountain No. 3, contributing to the settlement. You can see stone separations due to foundation settlement, and considerable water staining inside the Royal Mausoleum. Look to the south through the Enigmatic Window for a view of the Guardhouse beyond (Figure 4-8).

To the Royal Residence (Conjunto 3)

Several steps up and to the right you will see the entrance to the Royal Residence. Just inside, a large horizontal stone ring protrudes from the wall about 1.5 meters (5 feet) off the floor (Figure 10-22). This ring was carved from the rock formation that starts on the other side of the wall, showing an innovative use of a natural feature. Feel the smoothness of the inside of the ring. Perhaps a puma was tethered here, or it may have been a receptacle for a torch or staff.

FIGURE 10-22▲
The Inca may have used the horizontal ring in the Royal Residence for tethering a llama or puma, or to hold a torch or royal staff.

FIGURE 10-23►
The largest of the ceremonial sites at Machu Picchu is the Sacred Plaza bordered by the Principal Temple, the Temple of Three Windows, a utilitarian building on the far side, and the viewing platform on the west. The Priest's House is in the lower right of the photograph.

On the map, look for Room 5, which was likely a "bathroom" with its own drainage system through the wall (Figure 5-16). Room 7 has been identified as a cooking area because blackened shards were found there. Various other rooms supported the household.

To the Rock Quarry

Climb up the Stairway of the Fountains to the Quarry. The small circular foundations were likely those of huts used by the stoneworkers. Notice the rock on which an archaeologist experimented with splitting a rock using the ancient Egyptian method (Figure 8-23)—do not mistake it as an example of Inca stone masonry! The Inca used hammer stones, hundreds of which Hiram Bingham and others found in this area.

Arriving at the Sacred Plaza (Conjunto 4)

The next stop is the Sacred Plaza, the largest of the ceremonial sites with buildings on three sides (Figure 10-23). On the fourth open side you can look down the precipitous slope to the Urubamba River below and across the valley to the San Miguel Mountains. A semicircular foundation wall also juts out on this side, and is said to face the same direction as the grand semicircular wall of the Temple of the Sun in Cusco.

To the east, the Temple of the Three Windows looks down upon the Main Plaza, the eastern urban sector, and the mountains beyond (Figure 10-24). Exquisitely fitted polygonal (many-

FIGURE 10-24▼
Huge polygonal stones form the Temple of Three Windows. Similar large stones provide a foundation for the heavy walls.

◄FIGURE 10-25
From below the Temple of Three Windows, one can fully appreciate the great pieces of granite rock that were carefully shaped and fitted, even into the sturdy foundation.

sided) stones of awesome size and shape make up its walls and windows (Figure 10-25). Originally there were five windows; the two outer ones were closed off to form niches instead. The windows are much larger than any others in Machu Picchu. The triangular gables at both ends and the protruding round pegs show that this temple was roofed. A wooden beam would have spanned the front, supported by cuts in the side walls and the central upright stone (Figure 8-14).

Through the Intiwatana (Conjunto 5)

Check out the Priest's House (No. 4), which has fine stonework and a stone bench outfitted with drainage channels (Figure 10-26). The stairway up to the Intiwatana announces that you are approaching a site of major significance (Figure 10-27). As you leave the Priest's House, turn right for one set of stairs, then a ramp, and another set of stairs. At the head of these stairs is another enigmatic horizontal ring cut from the natural rock—but much smaller than the one in the Royal Residence. It, too, could have held a torch or a royal staff. The steps continue up this steep natural pyramid to a platform with a three-sided building on the West Side. Here, changes made to the original building are apparent. The door on one side has been filled in to create a window (Figure 8-18). On the east side of the platform are Machu Picchu's best "image stones." They were carved to represent the triangular peaks of Mount Yanantin in the far background and the

FIGURE 10-26▼
A stone bench in the Priest's House has small channels to drain water and provide a dry seat. The trapezoidal niches behind are geometrically spaced.

◄FIGURE 10-27
The wide granite stairway leading from the Priest's House to the Intiwatana announces the importance of the top of the pyramid.

FIGURE 10-28▲
The Intiwatana stone is 2 meters (7 feet) high and cut from the living rock. The stonecutters removed a portion of the stone to create the smooth surface. In some ways, the Intiwatana tends to reflect the shape of the peaks behind.

heavily forested, rounded Putucusi Mountain on the other side of the Urubamba River (Figure 2-11).

The final platform, the highest point in the royal estate, is the site of the famous Intiwatana stone (Figure 10-28). To your left is an arrow stone aimed south at Salcantay Mountain (Figure 1-12). The rock that formed the peak of this natural pyramid was sculpted down to create this piece of exceptional elegance and beauty (Figure 1-3). Bingham gave it the name "Intiwatana," a Quechua Indian word meaning "the place to which the sun was tied," hence the term "hitching post of the sun." It was neither an observatory nor a sundial.

From this high point you have a breathtaking 360-degree view. To the south is Machu Picchu Mountain, to the west is the San Miguel Range, and to the north is Huayna Picchu. The entire eastern sector lies to the east of you. From left to right: the *wayronas* of the Sacred Rock; the curious, jagged, upright rocks on a huge platform; the wall with the three symmetrical double-jamb doorways; a grand staircase; the Artisans' Wall; and the Temple of the Condor.

Seeing the Sacred Rock (Conjunto 6), the Unfinished Temple, and the Petroglyph

Head down the stairway on the north side of the Intiwatana hill (Figure 5-8), down to and across the Main Plaza to the Sacred Rock on the north end of the eastern sector. The rock is set on a pedestal with its

own small plaza flanked by two matching *wayronas* (Figure 10-29). The rock may have been cut to symbolize the mountains behind it (Figure 2-14).

The area has been restored by Dr. Alfredo Valencia Zegarra. The details show how carefully the ancient foundations were designed and constructed to prepare a solid foundation for the walls. This is also an opportunity to examine the details of a typical thatched roof (reconstructed). A large rock with a drip line cut into it is behind the right (south) *wayrona* (Figure 5-17). This drip line indicates that in Inca times the roof thatch was much thicker so the rain would drip off the roof and be carried away by the groove in the rock.

FIGURE 10-29▲
The Sacred Rock with its two wayronas *is at the north edge of Machu Picchu. This view is from the summit of Huayna Picchu via a 300-mm lens.*

To the south of the Sacred Rock lies the terraced mound with many huge, jagged rocks jutting upward. You saw this area from the Intiwatana. Here also is a petroglyph about 14 inches in diameter incised on a small rock. It lies horizontally and looks fairly insignificant. However, it is Inca and was recorded by Bingham. The clearest lines form four right angles, with each space containing smaller lines (Figure 10-30).

FIGURE 10-30▼
A petroglyph is carved into a flat rock on the terrace of the Unfinished Temple south of the Sacred Rock. Bingham photographed this petroglyph during his 1912 expedition.

A hike on top of the mound, with its strange, capricious rocks and some cut-but-unfinished stones, evokes feelings of discovery. Not many people come here, nor has it been excavated. It is thought to be an unfinished temple.

Visiting Conjuntos 7 and 8 And The Unfinished Wall

Head south on a somewhat rocky but very manageable trail that starts just behind the Sacred Rock on the east side of peripheral area C. The retaining wall along this trail is made of huge unfinished boulders. The rocks were placed first, then would have been hammered down to a straighter and smoother surface. A short walk takes you to the back entrance of Conjunto 7. This enclosure had some status, as evidenced

FIGURE 10-31► *Two mortar-like stones are the focus of this well-built structure after which the Group of the Mortars was named.*

FIGURE 10-32▼ *A Bingham 1912 photograph from the* National Geographic Magazine *archives shows a boy holding what might have been a grinding stone found near the mortars. Compare this picture with Figure 10-31.*

by its double-jamb doorway at the front entrance. Angle through the *conjunto* and exit through the double-jamb doorway overlooking the Main Plaza. Turn right to look at the massive wall (this is the south wall of peripheral area C). Notice what appears to be the figure of a bird about 5 meters (15 feet) long: beak, head, body and tail feathers shaped by the placement of stones (Figure 8-21). The perceived figure is considered random stone placement.

To the Three Doorways (Conjunto 9) and the Artisans Wall

It is a short distance from the entrance of Conjunto 8 to the service entrance of Conjunto 9. As you can see from the floor plan, it has three units of unusual symmetry with matching walls, doorways, and even niches. Each of these units, while not completely separated from each other, has a double-jamb doorway to the path overlooking the Main Plaza.

A grand staircase takes you down to the Artisans Wall, featuring some of the finest stonework in Machu Picchu (Figure 8-1). Instead of the coursed masonry of "the most beautiful wall," here each stone has a different shape, cut and fitted with utmost care and intricacy, forming an exquisite mosaic. And instead of gray, the wall has a warm, salmon-colored tone. The drain holes that are built into the wall are another indication of the meticulous engineering that produced this remarkable site (Figure 5-10).

◄FIGURE 10-33
The staircase leading to Conjunto 14 is cut from the living rock, as is the banister on the left.

FIGURE 10-34▼
From Conjunto 14, a grand staircase curves up to Conjunto 15.

Group of the Mortars (Conjuntos 14, 15, and 16)

Follow the Artisans Wall and take the left ramp heading east up to Conjuntos 14, 15, and 16. The entrance is a double-jamb doorway on the left, though only the lower portion remains. From the inside you can see how the barholds were integrated into the walls. The barhold on the left side has the sides, bottom, and bar all carved from one rock.

The first, Conjunto 16, is called the Group of the Mortars because of the two "mortars" (as in mortar and pestle for grinding grain) carved in rocks on the floor of the building on the immediate right (Figure 10-31). A Bingham photograph shows what could have been a rocker (large pestle) found near the "mortars" (Figure 10-32). However, the purpose and significance of the building and its stones is still being debated.

Continue through the second double-jamb doorway. Notice that the barholds on either side are perpendicular to each other rather than straight across and, in fact, one of them has no bar. This suggests that a pole or other stiff object was placed in the bar-less opening and tied to the bar on the other side of the doorway.

Continue north up a staircase with a banister, cut ingeniously into the living rock, and into Conjunto 14 (Figure 10-33). There a large patio and a grand staircase curve up to Conjunto 15 (Figure 10-34). Near the top of the staircase is a small opening with a large cavern below.

FIGURE 10-35▲
This 12-entrance kallanka *in Conjunto 10 is the largest single building in Machu Picchu and has a spectacular view of the triangular Yanantin Mountain.*

Sometimes you can feel air flowing out. Another such hole is at the top of the staircase just around the corner. One can only imagine the foundation work required to build this higher level of structure on such a jumble of rocks. Here is another shrine facing east—a large, carved natural rock with partial walls setting it off.

Going back down the staircase and turning north (left), walk between some buildings (Conjunto 14) and exit through a narrow opening.

About Conjuntos 10 and 11

Conjunto 11 consists of six, two-storied symmetrical *qolqas* (storehouses), conveniently located along the staircase that led both up to the elite residential area (Conjunto 9) and down to the lower agricultural terraces (see Chapter 6). Proceed up the stairway on the north side to see how easily both the lower and upper floors could be accessed. Windows in the gables of the second story were part of a system designed to help preserve the agricultural products. These stairs also take you to Conjunto 10, immediately to the north, which has the largest single building in Machu Picchu. With 12 entrances, it was another *kallanka,* like the one near the Guardhouse (Figure 10-35).

To Conjunto 12

Go back down the stairs and to the south to Conjunto 12, immediately to the east of Conjunto 11 and overlooking the Urubamba Valley. The dominating feature of this small *conjunto* is a beautifully carved natural rock with a seat and, on its edge, three small protrusions that seem to represent mountains and valleys in miniature (Figure 2-12). A cave, a maze of small rooms leading from one to another, and a final small platform with a panoramic view are at the lower level (take the stairs on the south side).

FIGURE 10-36▼
The entrance to the solar observatory of Intimachay (Cave of the Sun) is to the right of the wall. The window admits the sun's rays for a moment a few days each year to illuminate the back wall of the cave.

Tour Intimachay

The Intimachay (Cave of the Sun) is a solar observatory for which you can allocate between 20 minutes and one hour. Take the path between Rooms 4 and 5 of Conjunto 13 and continue south. Just before the path turns right at the corner of Conjunto 16, take a sharp left turn to a short staircase that takes you down to the Intimachay (Figure 10-36), named by Archaeoastronomer David S. Dearborn (Bauer and Dearborn 1995). Dearborn's investigations in 1984 concluded that the cave, altered and embellished by the Inca, was designed to admit the light of the rising sun through the window of the cave for only a brief period around the December solstice. The rays of the rising sun, peeking over a dip in the ridge of the mountain about 3 kilometers (2 miles) away, would pierce through the window and below the hanging rock to the back wall only during that brief time of the year.

Adjacent to the Intimachay, immediately to the north, is a large overhang. Notice the carefully engineered cutaway that directs rainwater to drip straight down rather than move along the underside of the overhang.

To the Temple of the Condor (Conjunto 17)

You can access The Temple of the Condor from different directions and doorways. If you happen to be at the lowest of the fountains (Fountain No. 16), go up the steep staircase on the east side to the upper double-jamb door. If you come from the Artisans Wall, go down the ramp to the east, to the second entrance on the right (a double-jamb doorway). Avoid the first doorway because it is definitely the back door and may not even have been a doorway in Inca times. If you have just been to Intimachay, you will be coming from above. You will be passing a large, white slanting natural rock with a platform cut into its upper end, now called The Slide for obvious reasons. Its purpose in Inca times is unknown. Take the stairway below the Slide, then up a few stairs to the left to the double-jamb doorway. This is the most dramatic entrance.

You are about to enter a most fascinating and intricate place (Figure 2-13). Here, the Inca engineers outdid themselves in using and embellishing the natural formations. The temple is complex, with many levels and subterranean caves. This is a great place to explore.

The condor is South America's largest bird with a wing span of up to 8 feet and capable of flying to great heights. Even today, Andean people revere it as a symbol of power and majesty. You can go under the left "wing," which leads to a cave tall enough to stand in, with niches and an entrance on the south side. Or, you can enter that cave directly through that south entrance by taking the stairs up and around the south side. There you will find unexpected views through cavernous openings. Below the right "wing" is a small opening with a few steps leading downward. Excavations have discovered man-made subterranean passageways that can only be accessed by crawling and are now closed off.

On the south side is a two-storied building with an interior stairway, also inaccessible. The first floor opens to a terrace and leads to steps down to Fountain No. 16. This is the only fountain that is pri-

vate since it can be accessed only from the Condor Conjunto. For additional privacy the water enters the fountain at ground level on the public side of the wall and then makes double turns before falling into the basin. You can go down to this fountain by going back out the double-jam doorway on the upper side of the *conjunto* and then going down the steep stairs on the east side of the *conjunto*.

Conjunto 18, located south of the Condor Conjunto, was another residential area.

Heading Back to the Hotel

Consult the map at this point. If you decided to go down to Fountain No. 16, you can work your way up the stairs, backtracking to Fountains 15, 14, 13, etc. You can continue up the Stairway of the Fountains to appreciate their commonality and their unique designs (see Chapter 4). The construction of Fountain No. 10 (Figure 2-9) is particularly unusual. The water is brought through the rock and around to fall into its basin from the opposite direction from all the other fountains. At Fountain No. 7 you join the main north–south path along the eastern edge of Conjuntos 2 and 3 (Temple of the Sun and the Royal Residence).

If instead you exited the Temple of the Condor on the north side, you can go up the ramp and over to the same main north–south path as above. Several side trips are described below.

Side Trips

Several side trips at Machu Picchu may be of special interest to civil engineers or discriminating tourists who have more time and may want to visit "backstage." For instance, if you are staying overnight at Machu Picchu, after your basic tour and lunch, take an afternoon hike to Intipunko and the Inca Bridge.

FIGURE 10-37▶ *High up near the top of Huayna Picchu, the Inca built terraces that seem to hang on the cliffside. The Urubamba River is seen below to the left.*

The following morning you can climb Huayna Picchu, which has a spectacular route and mountaintop ruins. Or, instead of Huayna Picchu, you might choose to visit the Temple of the Moon located on the north side of Huayna Picchu roughly 340 meters (1120 feet) below the summit.

If you are staying longer, we recommend you hike up Machu Picchu Mountain for grand sweeping views and glorious photographic opportunities.

The lower east flank of the Machu Picchu ridge is not accessible to the public at this time because of the rugged and hazardous terrain, poisonous snakes, ants, spiders, and rare fauna including the Bespectacled Bear. Someday in the future this area may be opened up for the hardy tourist who wants to climb up to Machu Picchu from the Urubamba River via the long-lost Inca Trail.

Hiking to Intipunko

A two-hour roundtrip hike to Intipunko (Gate of the Sun) is well worth the travel. Without a great deal of effort, you are treading on the famous Inca Trail from Cusco, just a tiny piece of the vast network of roads that connected the sprawling Inca Empire, an empire larger than that of Alexander the Great (Figure 9-12).

Intipunko was a final checkpoint where travelers on the Inca Trail were approved before heading down to Machu Picchu. It is ancient travelers and present-day trekkers' first view of Machu Picchu, and a glorious site it is. If the mists are swirling up from the Urubamba River, it looks like a city in the clouds (Figure 1-1). Inca ruins along the trail include a carved *huaca* (shrine) with buildings and a wide granite staircase at the shrine that disappears into the dense forest. Allow three hours for the trip if you want to absorb all the wonders along this trail and take photographs.

FIGURE 10-38▲
The descent trail from Huayna Picchu takes you through this imposing doorway of a special building. Machu Picchu can be seen below.

Inca Drawbridge

The trip to the Inca drawbridge is the shortest and easiest side trip. At peripheral area F (the Terrace of the Ceremonial Rock and Guardhouse) you can see the trail on the map along one of the terraces and a series of steps to the edge of the map with a red arrow pointing to Inca Trail to Drawbridge. You simply follow this easy trail for about 20 minutes. In some places the drop-off on your right is steep because you are following the face of a cliff. This becomes very obvious when you see the "bridge"—an ingenious rock wall built out from a very steep cliff, with a strategic gap (Figures 2-4 and 10-8). Timbers span the gap now, as in the past. The timbers could simply be removed to make this route to Machu Picchu inaccessible. Take the barrier sign seriously and do not venture beyond it. Come here on a clear day or the bridge may be veiled in mist.

Climbing Huayna Picchu

A climb to the top of Huayna Picchu is an adventure, as you are literally walking in the footsteps of the Inca. The well-developed trail to the summit includes many remarkable granite staircases. Ruins along the trail are fascinating (Figure 2-15), intermediate views are inspiring, and the high elevation stone terrace walls are impressive (Figure 10-37).

Near the summit, the trail splits with the right fork leading to a tunnel that ends in a series of stone, hand-hewn carved steps. The left fork, via a long, steep stairway, leads to a small building with an impressive doorway (Figure 10-38). The fork in the trail means that one can make the final elevation gain one way and descend the other. We recommend the tunnel route for the ascent because the other route provides great views on descent.

FIGURE 10-39▶
A lookout platform, or perhaps a platform on which to set a survey marker stone, lies along the trail near the summit of Huayna Picchu.

FIGURE 10-40▲
Looking south from the summit of Huayna Picchu, sometimes one can see Salcantay Mountain beyond the skyline. An arrow stone at the summit points toward this holy mountain.

The trail to the Huayna Picchu summit begins just north of the Sacred Rock where one must sign out with the guard prior to an early afternoon cutoff time. Do not miss the sign out—it is for your safety. A well-defined path with granite stairways heads for the narrow saddle between the two peaks of Huayna Picchu and the smaller Uña Picchu. After about 10 minutes, at the far end of the saddle, the path reaches the actual side of Huayna Picchu. The steep climbing begins here—sometimes on a rocky trail, sometimes on earth, and sometimes on formal stairways. Take care on the trail and do not become complacent, because a slip could prove fatal.

After about 40 minutes on the trail you will arrive at a switchback to the right where the first of the high-level terraces comes into view. When you reach the stone walls of the terraces, stop a moment to admire the incredible stone workmanship and the granite stairs. The view unfolds a few minutes after the fork in the trail, where you arrive at a viewing platform overlooking Machu Picchu and a structure below (Figure 10-39).

If you took the "tunnel route," when you reach the summit of Huayna Picchu (after about one hour), you will encounter a jumble of huge boulders. Look around, because you are now at the top of an Inca holy mountain with an incomparable view (Figure 10-40).

After relishing the bird's-eye view of Machu Picchu with all of its many details, let your eyes single out the Inca Trail at the Guardhouse.

◄FIGURE 10-41
The Temple of the Moon, on the north side of Huayna Picchu Mountain, is built under a huge rock overhang, much like a cave. There are no rooms, but only walls, niches, and recesses.

Visually follow the trail up to the ridge to Intipunko and then look southerly to see the summit of Salcantay. If you are not certain which direction is south, look for the V-shaped stone with a "seat" on the summit of Huayna Picchu which points south to Salcantay, often cloud-covered. Salcantay lies south, 8 degrees east from Huayna Picchu, a distance of about 20 kilometers (12 miles).

The elaborate platforms, the small building through whose door the alternate trail passes, the large slide rock, and the impressive terraces all attest that this was a holy mountain to the Inca along with Machu Picchu Mountain and Putucusi—the three of which protected Machu Picchu. Even today, the Quechua Indians revere the three mountains.

While the hike can be completed in about 1 1/2 hours (one hour up and l/2 hour down), plan for at least two to three hours to be able to appreciate the significance and beauty of the views from Huayna Picchu and its Inca remains along the trail. Finally, take care as you hike, because a fall can lead to catastrophe. Remember, the trail can be slippery if wet!

Temple of the Moon

You will find The Temple of the Moon on the north side of Huayna Picchu. Start at the same point you did for climbing Huayna Picchu and use the same trail. After a short hike and climb, a sign pointing to the left indicates that the temple is about one hour away. Do not use this trail if it is raining or if the trail is wet.

The Temple of the Moon is an unusual structure with fine stonework, built under a huge rock overhang or cave (Figure 10-41). The Inca did not create rooms here, just walls, recesses, and niches. Some of the niches are set off by flat double-jamb "doorways"—the only ones like this in the Machu Picchu area. A large rock has been carved

◄FIGURE 10-42
Beyond the Temple of the Moon is the tallest double-jamb doorway in the Machu Picchu area.

with a seat or "altar". The Inca revered mountains and the gods that inhabited them, and caves, like springs, were thought to be entrances for the gods.

Continue down the trail through the tallest double-jamb doorway in the Machu Picchu area (Figure 10-42). Note that the double-jamb is on the outside, i.e., coming up from below it would signify the approach to a special place. Additional structures located further down (Figure 10-43) include a small, walled cave with a doorway and crudely fashioned windows.

FIGURE 10-43▲
On the other side of the double-jamb doorway are a series of buildings with doorways, windows, niches, and a walled cave.

The Temple of the Moon was an important complex; it may have been a destination for worship from Machu Picchu, or perhaps it was another entrance to Machu Picchu up from the Urubamba River. On the return hike you will get a new vista of Machu Picchu from the northeast.

Hiking Machu Picchu Mountain

Lying south of Machu Picchu and 500 meters (1640 feet) higher is Machu Picchu Mountain (Figure 6-3). A two-hour hike to the summit provides breathtaking views along the trail and an incomparable view from the summit (Figures 1-5 and 2-6). Begin the trail by entering an opening through the terrace wall on your right about 150 meters (500 feet) up the trail from the Guardhouse, as shown on the map. The first portion of the trail is not so well-defined because of the thick forest beyond the terrace, but it later becomes self-evident. If you lose the trail, go back to the terrace and try again.

On the way up to the summit, be sure to note features along the way, such as a cave, the quarry of green rock lying in the Machu Picchu fault line, stone platforms for viewing the expansive sights in three directions, and the high quality granite stairway on the steep portions of the trail. As you near the summit, take particular note of the near-perfect sweeping, curved granite staircase.

The quality of the trail, the adjacent ruins, and the summit ruins of a small building, indicate that Machu Picchu Mountain, like Huayna Picchu, was more than a signal station and more than a place to watch for intruders. It was a holy mountain. Even today, the Quechua Indians respect Machu Picchu Mountain and fly a summit flag representing the Inca Empire (Figure 10-44). From the summit, note the following:

- The Machu Picchu fault running easterly from the mountain and through Aguas Calientes and beyond
- The rounded peak of the holy Putucusi across the Urubamba River
- Mount Veronica to the east, along with the rugged peaks of its range
- The tortuous configuration of the Urubamba River as it encircles Huayna Picchu
- The careful layout of Machu Picchu as it drapes over the ridge topography like a quilted blanket
- The snow-capped peak of Pumasillo Mountain to the west.

Also look for the summit of Salcantay, or at least where it should be, keeping in mind that it lies south from the summit.

The views from the summit of Machu Picchu Mountain are so extraordinary you will want to use a telephoto lens and a tripod to ensure utmost steadiness and increased sharpness of your photographs. The view from the trail is more dramatic while descending. Be sure to take along at least one bottle liter (1 quart) of water. It is also nice to have a lunch with you to allow for a leisurely journey.

FIGURE 10-44▲
Even today, the descendants of the Inca (Quechua Indians) fly their flag from the summit of Machu Picchu Mountain. Here, Wright is also flying the Explorer's Club flag.

References

Adams, C. 1975. Nutritive value of American foods. *Agr. Handbook* 456. Washington, DC: U.S. Department of Agriculture, p 296.

Allen, R.G., M. Smith, L.S. Pereira, and A. Perrier. 1994a. An update for the calculation of reference evapotranspiration. *International Commission on Irrigation and Drainage* 43(2): 35–92.

Allen, R.G., M.E. Jensen, J.L. Wright, and R.D. Burman. 1989. Operational estimates of reference evapotranspiration. *Agronomy Journal* 81: 650–62.

Allen, R.G., M. Smith, A. Perrier, and L.S. Pereira. 1994b. An update for the definition of reference evapotranspiration. *International Commission on Irrigation and Drainage* 43(2): 1–34.

American Society of Civil Engineers. 1992. Design and Construction of Urban Stormwater Systems. *ASCE Manuals and Reports of Engineering Practice No. 77*. New York: American Society of Civil Engineers.

———. 1969. *Stability and Performance of Slopes and Embankments*. New York: American Society of Civil Engineers.

American Water Works Association. 1969. *Water Treatment Plant Design*. New York: American Water Works Association.

Bauer, B.S. and D.S. Dearborn. 1995. *Astronomy and Empire in the Ancient Andes*. Austin, TX: University of Texas Press.

Bingham, H. 1913. In the wonderland of Peru. *National Geographic Magazine* (April 23): 387–574.

———. 1930. *Machu Picchu: A Citadel of the Incas*. New Haven, CT: Yale University Press.

Caillaux, Victor Carlotto. n.d. La Geologia en la Conservación del Santuario Histórico de Machu Picchu.

Diaz, H.F. and G.N. Kiladis. 1992. Atmospheric teleconnections associated with the extreme phase of the southern oscillation. In *El Niño Historical Paleoclimatic Aspects of Southern Oscillation*. Edited by H.F. Diaz and V. Markgraf. Cambridge, England: Cambridge University Press, p. 13.

Frost, P. 1989. *Exploring Cusco The Essential Guide to Peru's Most Famous Region*. Lima, Peru: Nuevas Imagenes.

Hemming, J. and E. Ranny. 1982. *Monuments of the Incas*. Albuquerque, NM: University of New Mexico Press (1990 reprint).

Jensen, M.E., R. D. Burman, and R.G. Allen. 1990. *Evapotranspiration and Irrigation Water Requirements Manual of Practice No. 70*. New York: American Society of Civil Engineers.

Kalafatovich, C.V. 1963. Geologia de la Ciudadela de Machupicchu y sus Alrededores. *Revista Universitaria No. 121*. Universidad Nacional del Cusco.

Knapp, C.L., T.L. Steffel, and S.D. Whitaker. 1990. *Insulation Data Manual*. SERI/SP-755-789. Golden, CO: Solar Energy Research Institute.

Kolata, A. 1993. *The Tiwanaku Portrait of an Andean Civilization*. Cambridge, MA: Blackwell.

Lee, V.R. 1988. The lost half of Inca architecture. Paper presented to the 28th Annual Meeting of the Institute of Andean Studies, January 8, 1988, Berkeley, CA.

Lee, V.R. with N.G. Lee. 1999. The Sisyphus project: moving big rocks up steep hills and into small places. Paper presented to the 39th Annual Meeting of the Institute of Andean Studies, January 8, 1999, Berkeley, CA.

Leverton, R.M. 1959. Recommended allowances. *Food, The Yearbook of Agriculture*, 1959. Washington, DC: U.S. Department of Agriculture.

MacLean, M. 1986. *Sacred Land Sacred Water: Inca Landscape Planning in the Cusco Area*. Ph.D. diss, University of California at Berkeley.

Marocco, René. 1977. Geologie des Andes Peruviennes: Un Segment W.E. de la Chaine des Andes Peruviennes; la Deflexion D'Abancy—Etude Geologique de la Cordillere Orientale et des Hauts Plateaux Entre Cusco et San Miguel. These, Academie de Montpellier Universite des Sciences et Techniques de languedoc.

Maurtua, V.M. 1906. *Juicio de Limites entre El Perú y Bolivia*. Madrid, Spain: Imprenta de Los Hijos de M.G. Hernández.

Mosley, M.E. 1992. *The Incas and Their Ancestors. The Archaeology of Peru*. London: Thames and Hudson.

National Geographic Data Center, National Oceanic and Atmospheric Administration. 1986. *Tropical Ice Core Paleoclimatic Records: Quelccaya Ice Cap, Peru A.D. 470 to 1984*. Columbus, OH: Ohio State University Byrd Polar Research Center (distributed by NGDC, Boulder, CO).

Pedde, L.D., W.E. Foote, L.F. Scott, D.L. King, and D.L. McGalliard. 1978. *Metric Manual*. Denver, CO: U.S. Government Printing Office.

Poma de Ayala, Don Felipe Huamán. 1978. *Letter to a King*. New York: E.P. Dutton.

Protzen, J.P. 1993. *Inca Architecture and Construction at Ollantaytambo*. New York: Oxford University Press.

Reinhard, J. 1991. *Machu Picchu: The Sacred Center*. Lima, Peru: Nuevas Imágines.

Rowe, J.H. 1990. Machu Picchu a la luz de documentos de siglo XVI. *Historica*, 14(1): 139-154.

———. 1997. Personal correspondence to the authors, January.

Smith, M. 1993. CLIMWAT for CROPWAT. *Food and Agriculture Organization of the United Nations Irrigation and Drainage Paper 49*.

Thompson, L.G., et al. 1984. Tropical glaciers: potential for ice core paleoclimatic reconstructions. *Journal of Geophysical Research* 89(D3): 4638–46.

———. 1985. A 1500-year record of tropical precipitation in ice cores from the Quelccaya ice cap, Peru. *Science* 229(September 6): 971–73.

———. 1995. Late glacial stage and holocene tropical ice core records from Huascarán, Peru. *Science* 269(July 7): 46–50.

Thompson, L.G., M. Davis, E. Mosley-Thompson, and K. Liu. 1988. Pre-Incan agricultural activity recorded in dust layers in two tropical ice cores. *Nature* 336(6220): 763–65.

Thompson, L.G., E. Mosley-Thompson, W. Dansgaard, and P.M. Grootes. 1986. The little ice age as recorded in the stratigraphy of the tropical Quelccaya ice cap. *Science* 234(4774): 361–64.

Thompson, L.G., E. Mosley-Thompson, and B.M. Morales. 1989. One-half millennia of tropical climate variability as recorded in the stratigraphy of the Quelccaya ice cap, Perú. *Geophysics Monograph* (55): 15–31.

Thompson, L.G., E. Mosley-Thompson, P.A. Thompson. 1992. Reconstruction interannual climate variability from tropical and subtropical ice-core records. In *El Niño Historical Paleoclimatic Aspects of the Southern Oscillation.* Edited by H.F. Diaz and V. Markgraf. Cambridge, England: Cambridge University Press.

Time-Life Books. 1992. *Incas: Lords of Gold and Glory*. Alexandria, VA: Time-Life Books.

Valencia Zegarra, A. 1977. *Excavaciones Arqueologicias en Machu Pijchu: Sector de la "Roca Sagrada."* Cusco, Peru: Instituto Nacional de Cultura.

Valencia Zegarra, A. and A. Gibaja Oviedo. 1992. *Machu Picchu: La Investigacíon y Conservacíon del Monumentos Arquelogico despues de Hiram Bingham.* Cuzco, Perú: Municipalidad del Qosqo.

Wright, K.R. 1996. The unseen Machu Picchu: a study by modern engineers. *South American Explorer* 46(Winter): 4–16.

———. 2000. Inca trail discovery at Machu Picchu. *South American Explorer,* Winter.

Wright, K.R. and A. Valencia Zegarra. 1999. Ancient Machu Picchu drainage engineering. *ASCE Irrigation and Drainage Journal* 125(6): 360–369.

Wright, K.R. and K.A. Loptien. 1999. Machu Picchu soil stewardship. *Erosion Control* 6(2): 60–67.

Wright, K.R., R.M. Wright, M.E. Jensen, and A. Valencia Zegarra. 1997a. Machu Picchu: ancient agricultural potential. *Applied Engineering in Agriculture* 13(1): 39–47.

Wright, K.R., G.D. Witt, and A. Valencia Zegarra. 1997b. Hydrogeology and paleohydrology of ancient Machu Picchu. *Ground Water* 35(4): 660–66.

Wright, K.R., J.M. Kelly, and A. Valencia Zegarra. 1997c. Machu Picchu: ancient hydraulic engineering. *Journal of Hydraulic Engineering* 123(10): 838–43.

Wright, K.R., S.A. Dracup, and J.M. Kelly. 1996. Climate variability impact on the water resources of ancient Andean civilizations. Presented at the North American Water and Environmental Congress, June 22-28, 1996, in Anaheim, CA.

Wright Water Engineers, Inc. 1996. Letter report to Dr. José Altamirano Vallenas of the Instituto Nacional de Cultura Departmental Cusco reporting findings of the soil sample testing at Colorado State University. (Results may be obtained directly from Dr. James Self, Colorado State University, Soil, Water and Plant Testing Laboratory, Room A319, Natural and Environmental Science Building, Fort Collins, CO 80523-1120).

Figure Credits

Nine figures of historical significance by Hiram Bingham and his associates (Figures 2-17, 3-6, 4-9, 5-3, 6-1, 10-12, 10-15, 10-18, and 10-32) appear through the courtesy of Yale University and the National Geographic Society, to whom the authors are indebted.

The eleven sketches by Don Felipe Huamán, Poma de Ayala, published in 1614, were provided by Dr. Gordon McEwan of Wagner College, Staten Island, who photographed the sketches (Figures 1-14, 6-8, 8-19, 9-1, 9-3, 9-4, 9-5, 9-7, 9-13, 9-15, and 9-20) while Curator at the Dumbarton Oaks Museum, who made them available for this ASCE Press publication.

The U.S. Department of Commerce provided the photograph of the Quelccaya Ice Cap (Figure 6-4) and the U.S. Defense Mapping Agency as a special courtesy generously provided an aerial photograph of Machu Picchu (Figure 9-17). Vincent Lee, a Wyoming architect and Andean explorer, provided Figure 8-9. Dr. Marvin Jensen of Ft. Collins, Colorado, provided Figures 6-6 and 6-7. Figure 8-10 is a composite figure based on the work of individuals as noted on the figure.

All other figures are credited to individual members of the Machu Picchu Paleohydrological Survey Project team members as follows:

Julia M. Johnson, professional photographer with Peaks and Places Photography of Boulder and Vail, Colorado. Figures 1-2, 3-1, 5-9, 7-5, 6-10, 7-16, 7-18, 8-14, 8-16, 8-21b, 8-23, 9-18, 10-1, 10-4, 10-7, 10-19, 10-24, 10-31, 10-33, 10-43.

Kurt Loptien, Graphics Designer, Wright Water Engineers, Inc. Figures 1-6, 3-2, 4-2, 8-20.

Grosvenor Merle-Smith, Co-Master & Huntsman of Bull Run Hunt, Keswick, Virginia. Figures 1-1, 1-13, 2-1, 2-4, 2-8, 2-12, 2-13, 2-16,

4-11, 5-1, 5-5, 5-10, 5-19, 7-4, 7-9m, 7-15, 7-19, 8-7, 8-17, 9-12, 10-8, 10-26, 10-40.

Patricia Pinson, Archivist and Librarian, Wright Water Engineers, Inc. Figure 1-7 and 9-8.

Gary Witt, Hydrogeologist, Wright Water Engineers, Inc. Figures 1-11, 6-3, and 8-4.

Kenneth R. Wright, President of Wright Water Engineers, Inc., and Wright Paleohydrological Institute. Figures 1-4, 1-5, 1-10, 2-6, 2-7, 2-9, 2-15, 3-3, 3-4, 3-5, 3-8, 3-9, 3-10, 3-11, 4-1, 4-3, 4-4, 4-5, 4-7, 4-12, 5-2, 5-6, 5-7, 5-8, 5-11, 5-13, 5-14, 5-15, 5-16, 5-17, 5-18, 6-2, 6-5, 6-9, 7-9 a-l, n and r, 7-10, 7-12, 7-13, 7-17, 8-3, 8-5, 8-6, 8-8, 8-11, 8-15, 8-21a, 8-22, 9-2, 9-6, 9-10, 9-11, 9-16, 9-19, 10-2, 10-3, 10-14, 10-17, 10-21, 10-27, 10-29, 10-30, 10-36, 10-37, 10-38, 10-39, 10-40, 10-44.

Ruth M. Wright, award-winning photographer and attorney. Vice President of Wright Paleohydrological Institute. Figures 1-3, 1-8, 1-9, 1-12, 2-2, 2-3, 2-5, 2-10, 2-11, 2-14, 3-7, 4-6, 4-8, 5-4, 5-12, 5-20, 6-10, 7-2, 7-3, 7-7, 7-8, 7-11, 7-13, 7-14, 8-1, 8-2, 8-12, 8-13, 8-18, 9-14, 10-5, 10-6, 10-9, 10-10, 10-11, 10-13, 10-16, 10-20, 10-22, 10-23, 10-25, 10-28, 10-34, 10-35, 10-41, 10-42.

Alfredo Valencia Zegarra, Ph.D., Professor, Department of Anthropology, Universidad de San Antonio Abad in Cusco, Peru. Figures 4-10, 7-6, and 9-9

Index

Numbers in **bold** indicate page includes an illustration. A figure is denoted by *f*; a table by *t*; and a map by *m*.